FORTSCHRITTE DER PHYSIK

Volume 34 · 1986

Number 11

PROGRESS OF PHYSICS

CONTENTS:

AKADEMIE-VERLAG · BERLIN

ISSN 0015-8208 Fortschr. Phys., Berlin 34 (1986) 11, 687–790

FORTSCHRITTE DER PHYSIK

VOLUME 34 1986 NUMBER 11

Fortschr. Phys. **34** (1986) 11, 687–751

On the 1-Loop Renormalization of the Electroweak Standard Model and its Application to Leptonic Processes

M. Böhm, H. Spiesberger[1])

Physikalisches Institut, Universität Würzburg, FRG,

W. Hollik

II. Institut für Theoretische Physik, Universität Hamburg, FRG

Abstract

A renormalization scheme for the electroweak standard model is presented in which the electric charge and the masses of the gauge bosons, Higgs particle and fermions are used as physical parameters. The photon is treated such that quantum electrodynamics is contained as a simple substructure. Field renormalization respecting the gauge symmetry gives finite propagators and vertex functions. The Ward identities between the Green functions of the unphysical sector allow a renormalization that maintains the simple pole structure of the propagators in the t'Hooft-Feynman gauge. We give a complete list of self energies and all renormalization constants also in the unphysical Higgs and ghost sector. Explicit results are given for the renormalized self energies, vertex functions and boxes that enter the evaluation of 1-loop radiative corrections to fermionic processes.

We calculate the 1-loop radiative corrections to purely leptonic reactions like μ decay, $\overset{(-)}{\nu}_\mu$e scattering and μ pair production in e^+e^- annihilation. A test of the standard model is performed by comparing these low energy data with the results of the $P\bar{P}$ collider experiments for the W and Z boson masses.

1. Introduction

The recent discovery of the W and Z bosons at the $P\bar{P}$ collider at CERN [1] with values for the masses of these particles very close to those predicted by the Glashow-Salam-Weinberg model [2] was an important step in establishing this model as a good candidate for the gauge theory of the electroweak interaction. But also the experiments with low momentum transfers ($|q^2| \ll M_W^2$) [3] and at e^+e^- storage rings [4] contribute to a steady improvement of the determination of the structure and parameters of the electromagnetic and weak interaction. The accuracy of these experiments has reached a level which requires the inclusion of radiative corrections for an adequate theoretical discussion. This will be even more the case when the e^+e^- machines with energies up to 100 GeV, which are dedicated for the investigation of the detailed properties of the electroweak bosons, go into operation [5].

The standard model is a non-Abelian gauge theory of the electroweak interaction where the masses of the particles are generated with help of the Higgs mechanism. The

[1]) supported by the Deutsche Forschungsgemeinschaft.

renormalizability of quantum field theories of this class was proved already in 1971 by 'T Hooft [6]. This means that those parts occuring in the evaluation of Feynman diagrams of higher order which without regularization would become ultraviolet divergent can be absorbed by renormalization of the fields and couplings. The importance of the renormalization constants is not only to absorb divergences but also to complete the definition of the quantized field theory. The finite parts of the renormalization constants — fixed by the renormalization conditions — influence the results of the calculation of radiative corrections and therefore of physically observable effects.

Electroweak theories contain much more fields and parameters than quantum electrodynamics, moreover their structure is more complicated because of their non-Abelian, non-simple, spontaneously broken gauge symmetry. The choice of the renormalized parameters and their definition via measurable quantities as well as the definition of the weak mixing angle is not unique beyond the tree level.

Consequently several different schemes have been proposed in the literature [7—17]. The greater part [10, 12—55] deals with processes where $|q^2| \ll M_W^2$, like μ decay and ν scattering; refs. [9, 11, 16, 17] consider high q^2 e^+e^- annihilation. An attempt for their characterization can be made using the following criteria:

- Schemes with and without field renormalization; in the latter case S-matrix elements but not the Green functions such as self energies and 3-point vertex functions are finite.
- Field renormalization respecting the original gauge symmetry or not; Green functions are finite but have complicated properties under gauge transformations in the latter case.
- Determination of the parameters from low energy experiments like μ decay and $\overset{(-)}{\nu}_\mu e$ scattering or from high energy experiments i.e. measurements of the W, Z masses.

Of course, not all papers on electroweak radiative corrections fit simply into one of these categories.

In this paper we present a renormalization scheme for the standard electroweak model which is defined by the following conditions:

i) The physical parameters are the electric charge e, the masses of the W and Z bosons, the Higgs mass and the fermion masses. This set was introduced by Sirlin [10] and later used also by other authors [11, 15]. e is defined as the strength of the electromagnetic coupling in the Thomson limit, the masses as the position of the poles of the renormalized propagators. These parameters are directly accessible to experiment, since only the measurement of the fine structure constant α and of masses is required. The determination of masses in direct resonance production experiments is only very little influenced by radiative corrections. Bare masses and couplings do not occur; this avoids possible confusions in calculating cross sections in higher orders. The weak mixing angle θ_W and the Fermi constant G_F are no fundamental parameters. $c_W = \cos\theta_W = M_W/M_Z$, $s_W = \sqrt{1 - c_W^2}$ are only shorthand notations to simplify the formulas.

ii) Real photons couple to the electron without any admixture of Z^0 contributions. Therefore the QED subpart of the model is realized in a simple way. Consequently photonic radiative corrections can be treated separately; especially for $e^+e^- \to f\bar{f}$ they can be taken over from pure QED calculations [18, 19].

iii) Complete field renormalization respecting gauge invariance and the use of the 't Hooft-Feynman gauge lead to UV finite renormalized Green functions reflecting the gauge symmetry structure. We investigate the restrictions of the Slavnov-Taylor identities for the renormalization of the unphysical Green functions. Especially we perform a renormalization of the gauge fixing parameters in such a way that their poles are situated at M_Z^2, M_W^2, 0.

We add some comments on the relation of this scheme to previous work by other authors.

The on-shell scheme (i) with e and the particle masses as parameters has been widely used for various applications in the last years [8—12, 15]. Counter terms in the physical sector and amplitudes for scattering between spin 1/2 particles are presented in ref. [8] in a unitary gauge calculation.

A vanishing renormalized photon-Z^0 mixing for on-shell photons (ii), which allows to be as close as possible to QED, has also been used by the authors of ref.s [8, 11, 12, 15]. The treatment of field renormalization in [11, 15] differs from [8, 12] and ours, since more renormalization constants than symmetry multiplets are introduced. For physical S-matrix elements the results should be equivalent. This is also the case for the scheme without field renormalization [10].

The method of SAKAKIBARA [12] to generate counter terms and his renormalization conditions are nearly identical to ours in the physical sector; for the unphysical gauge boson, Higgs and ghost parts ref. [12] contains less counter terms since the gauge fixing part is not renormalized, according to Ross and TAYLOR [12]. One consequence of Sakakibara's procedure is that the relation $\sin^2 \theta_W = 1 - M_W^2/M_Z^2$ is no longer valid in higher order. This should not affect physical results if M_W, M_Z are rigorously used and auxiliary quantities avoided in final results. Since the evaluation and renormalization of all longitudinal gauge boson, Higgs and ghost self energies is not performed in [12] an explicit check of the equivalence of the pole structure also in the unphysical sector is not possible so far. However, Ross and TAYLOR [12] claim that in this scheme which has no gauge parameter renormalization the poles of the individual unphysical propagators are not the corresponding tree levels poles.

The intention of this paper is to give a self contained and elaborate discussion of the standard model renormalization and to provide the basis for the calculations of radiative corrections, in particular to e^+e^- annihilation, deep inelastic scattering and $P\bar{P}$ annihilation with $|q^2| \simeq M_W^2$. The paper is organized as follows:
Sect. 2 of this paper contains the definition of the complete Lagrangian and its parameters, sect. 3 the discussion of the Slavnov-Taylor resp. Ward identities, sect. 4 the renormalization conditions. The complete list of the Feynman rules including the counter terms can be found in the appendix A. In sect. 5 we list the 1-loop formulas for the unrenormalized self energies, the fermion gauge boson vertex functions and the renormalization constants. We give also all the unphysical Higgs and ghost self energies together with the renormalization constants which have not yet been presented so far. Sect. 6 contains the renormalized boson self and mixing energies and simple formulas for the renormalized gauge boson fermion vertices for arbitrary momentum transfer. Numerical results are shown for those self energies, vertices and box diagrams that enter the radiative corrections to electroweak processes between fermions. In the last section 7 we apply this renormalization scheme to the purely leptonic reactions μ decay, $\overset{(-)}{\nu}_\mu e$ scattering and lepton pair production in e^+e^--annihilation. A test of the standard model is performed using as input the experimental data for these processes together with the measured values for M_W and M_Z [1].

2. The Renormalized Lagrangian and the Feynman Rules of the Standard Electroweak Model

2.1 The classical Lagrangian, parameter and fields

Gauge theories of the electroweak interaction are constructed in such a way that at low energies and in lowest order the experimentally successful Fermi model is recovered. In the case of the standard model [2] the universality of the weak interaction is realized in the form of the gauge group $SU(2) \times U(1)$. The gauge symmetry is spontaneously broken with help of a minimal Higgs mechanism with a $SU(2)$ doublet of scalar fields such that the electromagnetic gauge invariance $U(1)^{em}$ is maintained. The standard model allows to predict from low energy experiments the masses M_W, M_Z of the heavy gauge bosons $W^\pm$, Z. The existence and main properties of these particles have recently been confirmed by experiments at the $P\bar{P}$ collider [1].

The classical Lagrangian of the standard model $\mathcal{L}_C$ is composed of the gauge, Higgs and fermion part:

$$\mathcal{L}_C = \mathcal{L}_{YM} + \mathcal{L}_H + \mathcal{L}_F. \tag{2.1}$$

According to the gauge group $SU(2) \times U(1)$ we have an isotriplet $W_\mu^a(x)$ and an isosinglet $B_\mu(x)$ of gauge fields with gauge coupling constants g_2 and g_1 leading to the Yang-Mills Lagrangian:

$$\mathcal{L}_{YM} = -\frac{1}{4}(\partial_\mu W_\nu^a - \partial_\nu W_\mu^a + g_2\varepsilon^{abc}W_\mu^b W_\nu^c)^2 - \frac{1}{4}(\partial_\mu B_\nu - \partial_\nu B_\mu)^2. \tag{2.2}$$

The complex Higgs doublet $\varphi(x)$

$$\varphi(x) = \begin{pmatrix}\varphi^+(x)\\ \varphi^0(x)\end{pmatrix} = \begin{pmatrix}\varphi^+(x)\\ (v + \eta(x) + i\chi(x))/\sqrt{2}\end{pmatrix}$$

with hypercharge $Y = 1$ is coupled to the gauge bosons and has a self coupling:

$$\mathcal{L}_H = (D_\mu\varphi)^+ (D^\mu\varphi) - \frac{\lambda}{4}(|\varphi|^2)^2 + \mu^2 |\varphi|^2 \tag{2.3}$$

with the covariant derivative:

$$D_\mu = \partial_\mu - ig_2 I^a W_\mu^a + ig_1\frac{Y}{2}B_\mu. \tag{2.4}$$

The left-handed fermion fields $\psi_{i\sigma}^L(x)$ are grouped into doublets (i = doublet index, σ = component of the doublet) of the weak isospin, the right-handed fields $\psi_{i\sigma}^R(x)$ into singlets, the hypercharges respecting the Gell-Mann Nishijima relation $Q = I^3 + Y/2$. The Lagrangian $\mathcal{L}_F$ which describes the interaction between the fermions, the gauge fields and the scalars then has the form[2]):

$$\begin{aligned}\mathcal{L}_F = \sum_i \{&\bar{\psi}_{i\sigma}^L i\gamma^\mu D_\mu^{\sigma\sigma'}\psi_{i\sigma'}^L + \bar{\psi}_{i\sigma}^R i\gamma^\mu D_\mu^{\sigma\sigma'}\psi_{i\sigma'}^R\\ &+ (-g_{i+}\bar{\psi}_{i+}^R\varphi^0\psi_{i+}^L - g_{i-}\bar{\psi}_{i-}^R\varphi^{0*}\psi_{i-}^L + g_{i+}\bar{\psi}_{i+}^R\varphi^+\psi_{i-}^L - g_{i-}\bar{\psi}_{i-}^R\varphi^-\psi_{i+}^L + \text{h.c.})\}.\end{aligned} \tag{2.5}$$

[2]) We do not write explicitly colour indices and the Cabbibo transformation of the quark fields i.e. we assume the coupling matrix $g_{ij} = g_i\delta_{ij}$ to be diagonal.

This completes the construction of $\mathscr{L}_C$ in terms of the fields W_μ^a, B_μ, φ, ψ^L, ψ^R and the parameters

$$g_2, g_1, \lambda, \mu^2, g_{i\sigma}. \tag{2.6}$$

$\mathscr{L}_C$ is invariant under local transformations of the group $SU(2) \times U(1)$:

$$g = \exp\left[ig_2 I^a\theta_a(x) - i\frac{g_1}{2}\, Y\theta_Y(x)\right],$$

namely

$$\frac{\delta \mathscr{L}_C}{\delta\theta^\alpha(x)} = 0, \qquad \alpha = \{a, Y\}. \tag{2.7}$$

A formulation where the physical content of the theory is more — but the symmetry less — transparent can be obtained by performing the following transformation of the gauge fields:

$$\begin{aligned} W_\mu^\pm &= (W_\mu^1 \mp iW_\mu^2)/\sqrt{2}, \\ Z_\mu &= \frac{M_W}{M_Z}\, W_\mu^3 + \left(1 - \frac{M_W^2}{M_Z^2}\right)^{1/2} B_\mu, \\ A_\mu &= -\left(1 - \frac{M_W^2}{M_Z^2}\right)^{1/2} W_\mu^3 + \frac{M_W}{M_Z} B_\mu \end{aligned} \tag{2.8}$$

and using the following parameters:

$$e, M_W, M_Z, M_H, m_{i\sigma} \tag{2.9}$$

with

$$\begin{aligned} e &= \frac{g_1 g_2}{(g_1^2 + g_2^2)^{1/2}}, \qquad M_W = g_2\mu/\sqrt{\lambda}, \quad M_Z = (g_1^2 + g_2^2)^{1/2}\, \mu/\sqrt{\lambda}, \\ M_H &= \sqrt{2}\,\mu, \qquad m_{i\sigma} = g_{i\sigma}\mu\,\sqrt{2/\lambda}. \end{aligned} \tag{2.10}$$

Each of the parameters (2.9) is directly accessible to experiments since for their determination measurements of the Thomson scattering cross section (for the electric charge e) and of the masses of the W boson, Z boson, Higgs boson and the fermions are required. This is the reason why we prefer the set of more physical fields (2.8) and parameters (2.9) as the basis for the formulation of the electroweak Lagrangian $\mathscr{L}_C$. It may be that for low energy processes the use of other parameters like the Fermi constant G_F and the weak mixing angle θ_W is more convenient [10, 12, 13]. The relation between M_W, M_Z and G_F, θ_W to lowest order is:

$$\begin{aligned} G_F/\sqrt{2} &= \pi\alpha/2M_W^2(1 - M_W^2/M_Z^2), \\ \cos\theta_W &= M_W/M_Z. \end{aligned} \tag{2.11}$$

Depending on the specific renormalization scheme some of the relations (2.8)—(2.11) get corrections from higher order contributions.

2.2 Gauge fixing and ghost fields

For the systematic treatment of the quantization of $\mathcal{L}_C$ and higher order calculations it is convenient to choose a renormalizable gauge. We introduce linear gauge fixings $F^\alpha(W, B, \varphi)$ of the 't Hooft type:

$$F^\pm = (\xi_1^W)^{-1/2}\,\partial^\mu W_\mu^\pm \mp iM_W(\xi_2^W)^{1/2}\,\varphi^\pm,$$
$$F^3 = (\xi_1^3)^{-1/2}\,\partial^\mu W_\mu^3 - M_W(\xi_2^3)^{1/2}\,\chi, \tag{2.12}$$
$$F^B = (\xi_1^B)^{-1/2}\,\partial^\mu B_\mu - (M_Z^2 - M_W^2)^{1/2}\,(\xi_2^B)^{1/2}\,\chi.$$

Then we add to $\mathcal{L}_C$ the term

$$\mathcal{L}_{\text{fix}} = -\frac{1}{2}\sum_\alpha (F^\alpha)^2 \tag{2.13}$$

and introduce the Faddeev-Popov ghost fields $u^\alpha(x)$ resp. $u^\pm(x)$, $u^Z(x)$, $u^\gamma(x)$ with the Lagrangian [21]:

$$\mathcal{L}_{FP} = \bar{u}^\alpha(x)\,\frac{\delta F^\alpha}{\delta\theta^\beta(x)}\,u^\beta(x) = \bar{u}^\alpha K^{\alpha\beta} u^\beta. \tag{2.14}$$

A particular choice of the gauge parameters ξ is:

$$\xi_1^W = \xi_2^W = \xi_1^3 = \xi_2^3 = \xi_1^B = \xi_2^B = 1. \tag{2.15}$$

This 't Hooft-Feynman gauge has the advantage that at least to lowest order the poles of the longitudinal parts of the gauge boson propagators, the unphysical Higgs fields $\varphi^\pm$, χ and the ghost fields are situated at M_W^2 or M_Z^2 and that no gauge field — Higgs field mixing occurs.

With $\mathcal{L}_{\text{fix}}$ and $\mathcal{L}_{\text{FP}}$ we have completed the construction of a renormalizable Lagrangian

$$\mathcal{L} = \mathcal{L}_C + \mathcal{L}_{\text{fix}} + \mathcal{L}_{FP} \tag{2.16}$$

for the standard electroweak model.

2.3 Multiplicative Renormalization

The Lagrangian (2.16) is the starting point for the calculation of Green functions and S matrix elements including radiative corrections. We renormalize not only the physical parameters but also the fields in order to arrive at Green functions that are finite. For S-matrix elements the results should of course be equivalent to those obtained without field renormalization (see e.g. Sirlin [10]). Since symmetry arguments were important in the construction of $\mathcal{L}_C$ we perform the multiplicative renormalization of $\mathcal{L}$ in such a way that the gauge symmetry is respected:

$$W_\mu^a \to (Z_2^W)^{1/2}\,W_\mu^a, \qquad B_\mu \to (Z_2^B)^{1/2}\,B_\mu,$$
$$\varphi \to (Z^\varphi)^{1/2}\,\varphi,$$
$$\psi_{i\sigma}^L \to (Z_L^i)^{1/2}\,\psi_{i\sigma}^L, \qquad \psi_{i\sigma}^R \to (Z_R^{i\sigma})^{1/2}\,\psi_{i\sigma}^R,$$
$$g_2 \to Z_1^W (Z_2^W)^{-3/2}\,g_2, \qquad g_1 \to Z_1^B (Z_2^B)^{-3/2}\,g_1, \tag{2.17}$$

$$\begin{aligned} &\lambda \to Z^\lambda (Z^\varphi)^{-2} \lambda, \quad \mu^2 \to (\mu^2 - \delta\mu^2) (Z^\varphi)^{-1}, \\ &v \to (Z^\varphi)^{1/2} (v - \delta v), \\ &g_{i\sigma} \to (Z^\varphi)^{-1/2} Z_1^{i\sigma} g_{i\sigma}, \\ &\xi_{1,2}^\alpha \to 1 + \delta\xi_{1,2}^\alpha, \qquad u^a \to \tilde{Z}^W u^a, \qquad u^B \to \tilde{Z}^B u^B, \qquad \bar{u}^\alpha \to \bar{u}^\alpha. \end{aligned} \tag{2.17}$$

These definitions of renormalized fields and parameters induce corresponding expressions for the fields (2.8) i.e. $W_\mu^\pm$, Z_μ, A_μ and the parameters (2.9). Writing

$$Z_i = 1 + \delta Z_i \tag{2.18}$$

we obtain $\mathcal{L} \to \mathcal{L} + \delta\mathcal{L}$ where the expression for $\mathcal{L}$ in the renormalized quantities is identical with the original one, but now contains the renormalized physical parameters and fields. The quantities δZ_i, δv, $\delta\mu^2$, $\delta\xi_i^\alpha$ occur in the counter term Lagrangian $\delta\mathcal{L}$. Their finite parts have to be fixed by the explicit renormalization conditions. Before doing this we study the restrictions which are imposed on the renormalization procedure by the Slavnov-Taylor identities of the theory.

The Feynman rules belonging to $\mathcal{L}$ and the counter terms from $\delta\mathcal{L}$ are listed in app. A.

3. Slavnov-Taylor Identities

3.1 The Becchi-Rouet-Stora transformation

The original gauge invariance of $\mathcal{L}_C$ is lost after the introduction of $\mathcal{L}_{\text{fix}}$ and $\mathcal{L}_{FP}$ but the complete Lagrangian $\mathcal{L}$ is invariant under gauge transformations involving also ghost fields $u^\alpha(x)$. In order to discuss this point we use the following condensed notation for the fields and their transformations:

$$\Phi_s = \{W_\mu^a(x), B_\mu(x), \varphi(x), \psi_{i\sigma}^L(x), \psi_{i\sigma}^R(x)\}, \tag{3.1}$$

$$\delta\Phi_s = (\Delta_s^\alpha + g^\alpha T_{st}^\alpha \Phi_t)\, \delta\theta^\alpha. \tag{3.2}$$

The inhomogeneous term Δ_s^α acts only on the gauge field part of Φ, T^α denotes the representation matrices of the $SU(2) \times U(1)$ generators. The transformation under which $\mathcal{L}$ is invariant — the Becchi-Rouet-Stora transformation [22] — is constructed in such a way that the parameters of the infinitesimal gauge transformation $\delta\theta^\alpha$ contain the ghost fields:

$$\delta\theta^\alpha(x) = u^\alpha(x) \cdot \bar{\lambda}. \tag{3.3}$$

($\bar{\lambda}$ is independent of α and has ghost number -1). Since (3.2) together with (3.3) defines a gauge transformation $\mathcal{L}_C$ is still invariant. The transformation of the ghost fields u^α, $\bar{u}^\alpha$ is defined in such a way that $\mathcal{L}_{\text{fix}} + \mathcal{L}_{FP}$ is also invariant:

$$\delta\bar{u}^\alpha = F^\alpha \cdot \bar{\lambda}, \tag{3.4}$$

$$\delta u^\alpha = -(K^{-1})^{\alpha\beta}\, \delta K^{\beta\gamma} u^\gamma = -\frac{1}{2} C^{\alpha\beta\gamma} u^\beta u^\gamma \bar{\lambda}, \tag{3.5}$$

where F^α is the linear gauge fixing operator (2.12), K the Faddeev-Popov kernel (2.14) and $C^{\alpha\beta\gamma}$ the structure constant of the gauge group.

3.2 The Slavnov-Taylor identities

The BRS symmetry of $\mathcal{L}$ induces symmetry relations between the Green functions of the theory. They can be derived in a compact form with the help of the path integral formalism. The generating functional W of the Green functions $\tau_{s_1 \dots s_n} = \langle 0 | \, T\Phi_{s_1} \cdots \Phi_{s_n} \, |0\rangle$:

$$\tau_{s_1 \dots s_n} = i^{-n} \left. \frac{\delta^n W}{\delta j_{s_1} \cdots \delta j_{s_n}} \right|_{j=0} \tag{3.6}$$

is defined by:

$$W[j, \omega, \overline{\omega}] = \int D\Phi_s D\overline{u} Du \exp\left\{i \int d^4x [\mathcal{L} + \Phi_s j_s + \overline{u}^\alpha \omega^\alpha + \overline{\omega}^\alpha u^\alpha]\right\}. \tag{3.7}$$

Here we have introduced sources j_s for the fields Φ_s ans sources ω^α, $\overline{\omega}^\alpha$ for the (anti)-ghost fields. From the invariance of $\mathcal{L}$, $D\Phi_s$ and $D\overline{u}Du$ under BRS transformations one obtains for W the identity:

$$\left\{ iF^\alpha \left[\frac{\delta}{i\delta j}\right] + j_t \left(\Delta_t{}^\beta + gT^\beta_{tt'} \frac{\delta}{i\delta j_{t'}}\right) \frac{\delta^2}{i\delta\omega^\alpha i\delta\overline{\omega}^\beta} \right\} W[j, \omega, \overline{\omega}]|_{\omega=\overline{\omega}=0} = 0. \tag{3.8}$$

From this Slavnov-Taylor identity [23] follow the desired relations between the Green functions τ by taking suitable derivatives with respect to the sources j_s and putting afterwards all $j_s = 0$.

A special class of relations, those which do not directly contain Green functions of ghost fields, results if the gauge fixing operator F^α is applied to eq. (3.8):

$$F^\alpha \left[\frac{1}{i} \frac{\delta}{\delta j}\right] F^\beta \left[\frac{1}{i} \frac{\delta}{\delta j}\right] W[j]|_{j=0} = i\delta^{\alpha\beta} W[0]. \tag{3.9}$$

This equation relates the gauge boson propagators $\Delta^{\alpha\beta}_{\mu\nu}(k)$ to the gauge boson Higgs mixing propagators $\Delta_\mu{}^{\alpha i}(k)$ and the unphysical Higgs propagators $\Delta^{ij}(k)$:

$$\begin{aligned}
&k^\mu k^\nu \Delta^W_{\mu\nu}(k) + 2M_W k^\mu \Delta_\mu{}^{W\varphi}(k) + M_W{}^2 \Delta^\varphi(k) = -i, \\
&k^\mu k^\nu \Delta^Z_{\mu\nu}(k) - 2iM_Z k^\mu \Delta_\mu{}^{Z\chi}(k) + M_Z{}^2 \Delta^\chi(k) = -i, \\
&k^\mu k^\nu \Delta^\gamma_{\mu\nu}(k) = -i, \\
&k^\mu k^\nu \Delta^{\gamma Z}_{\mu\nu}(k) - iM_Z k^\mu \Delta_\mu{}^{\gamma\chi}(k) = 0.
\end{aligned} \tag{3.10}$$

In order to get relations between self energies Σ we decompose the gauge field propagators into their transverse and longitudinal parts

$$\Delta^{\alpha\beta}_{\mu\nu}(k) = \left(-g_{\mu\nu} + \frac{k_\mu k_\nu}{k^2}\right) \Delta_T{}^{\alpha\beta}(k^2) - \frac{k_\mu k_\nu}{k^2} \Delta_L{}^{\alpha\beta}(k^2), \tag{3.11}$$

make use of the Lorentz covariance of the gauge Higgs mixing propagators

$$\Delta_\mu{}^{\alpha i}(k) = ik_\mu \Delta^{\alpha i}(k^2) = ik_\mu \frac{1}{k^2 - M_\alpha{}^2} \Sigma^{\alpha i}(k^2) \frac{1}{k^2 - M_i{}^2}, \tag{3.12}$$

split off the free parts of the propagators

$$\Delta^{\alpha\beta}_{T,L}(k^2) = \frac{i}{k^2 - M_\alpha{}^2} \left(\delta^{\alpha\beta} - \Sigma^{\alpha\beta}_{T,L}(k^2) \frac{1}{k^2 - M_\beta{}^2}\right) \tag{3.13}$$

and end with:

$$
\begin{aligned}
&k^2(\Sigma_L{}^W + 2M_W\Sigma^{W\varphi}) - M_W{}^2\Sigma^\varphi = 0,\\
&k^2(\Sigma_L{}^Z - 2iM_Z\Sigma^{Z\chi}) - M_Z{}^2\Sigma^\chi = 0,\\
&k^2\Sigma_L{}^\gamma = 0,\\
&k^2(\Sigma_L{}^{\gamma Z} - iM_Z\Sigma^{\gamma\chi}) = 0.
\end{aligned}
\tag{3.14}
$$

As a consequence of $U(1)^{\text{em}}$ gauge invariance the longitudinal photon self energy vanishes identically as is the case in pure QED.

By adding the appropriate counter terms (app. A) we arrive at identities for the renormalized self energies $\hat{\Sigma}$:

$$
\begin{aligned}
&k^2(\hat{\Sigma}_L{}^W + 2M_W\hat{\Sigma}^{W\varphi}) - M_W{}^2\hat{\Sigma}^\varphi = (k^2 - M_W{}^2)\,[k^2(\delta Z_2{}^W - \delta\xi_1{}^W)\\
&\qquad\qquad - M_W{}^2(\delta\xi_2{}^W + \delta Z^\varphi) - \delta M_W{}^2],\\
&k^2(\hat{\Sigma}_L{}^Z - 2iM_Z\hat{\Sigma}^{Z\chi}) - M_Z{}^2\hat{\Sigma}^\chi = (k^2 - M_Z{}^2)\,[k^2(\delta Z_2{}^Z - \delta\xi_1{}^Z)\\
&\qquad\qquad - M_Z{}^2(\delta\xi_2{}^Z + \delta Z^\varphi) - \delta M_Z{}^2],\\
&\hat{\Sigma}_L{}^\gamma = k^2(\delta Z_2{}^\gamma - \delta\xi_1{}^\gamma),\\
&(\hat{\Sigma}_L{}^{\gamma Z} - iM_Z\hat{\Sigma}^{\gamma\chi})\\
&= -k^2(\delta Z_2{}^{\gamma Z} - \delta\xi_1{}^{\gamma Z}) + M_Z{}^2\left(\delta Z_1{}^{\gamma Z} - \delta Z_2{}^{\gamma Z} + \frac{1}{2}\,\delta\xi_2{}^{\gamma Z} - \frac{1}{2}\,\delta\xi_1{}^{\gamma Z}\right).
\end{aligned}
\tag{3.15}
$$

As a consequence of these results the number of independent renormalization conditions for the unphysical propagators $\Delta^{\alpha\beta}_{\mu\nu}$, $\Delta_\mu{}^{\alpha i}$, Δ^{ij} is reduced. But eq. (3.15) is compatible with a renormalization where the poles of these propagators are located at $M_W{}^2$, $M_Z{}^2$, 0. This means that the structure which is realized in lowest order in the Feynman gauge can be maintained in all orders by a suitable renormalization of the gauge fixing parameters $\xi^\alpha_{1,2}$.

We do not work out the relations like (3.9) between the "unphysical" parts of higher Green functions since we do not need them for the investigation of the restrictions on the renormalization constants in the unphysical sector.

Let us have a look now at the ghost propagators. Differentiation of eq. (3.8) with respect to the sources of the gauge fields yields:

$$
\begin{aligned}
&[k^\mu\Delta^{\alpha\beta}_{\mu\nu}(k) + iM_\alpha\Delta_\nu{}^{\beta i}(k) + k_\nu G^{\alpha\beta}(k)]\,\delta^{(4)}(k - k')\\
&= -iC^{\gamma\beta\delta}\int d^4q\,G^{\alpha\gamma,\delta}{}_\nu(k, -k' + q, -q),
\end{aligned}
\tag{3.16}
$$

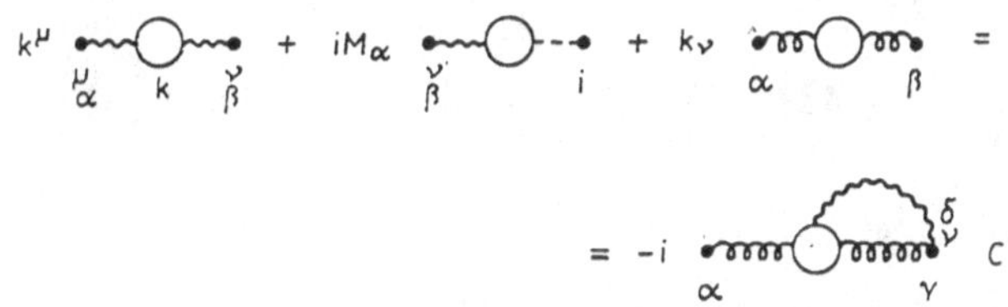

Here $G^{\alpha\beta}(k)$ denotes the ghost propagator and $G^{\alpha\beta,\delta}{}_\nu = \langle 0|\,(Tu^\alpha\bar{u}^\beta W_\nu{}^\delta)\,|0\rangle$ the gauge field – ghost three point function. Eq. (3.16) is in lowest order the usual relation between the longitudinal part of the gauge field propagator and the ghost propagators, in 1-loop order it allows to determine the ghost self energies from $\Sigma_L^{\alpha\beta}$, $\Sigma^{\alpha i}$ and the diagram on the r.h.s. An important consequence of eq. (3.16) is that renormalization in the ghost sector can be performed in such a way that the poles of the ghost propagators remain at M_W^2, M_Z^2, 0.

The identities (3.16) read in 1-loop order for the self energies:

$$\begin{aligned}
&\Sigma_L^{\gamma}(k^2) - \tilde{\Sigma}^{\gamma}(k^2) = \frac{\alpha}{4\pi} k^2 B_0(k^2; M_W, M_W), \\
&\Sigma_L^{\gamma Z}(k^2) - \tilde{\Sigma}^{Z\gamma}(k^2) = -\frac{\alpha}{4\pi}\frac{c_W}{s_W}(k^2 - M_Z^2)\, B_0(k^2; M_W, M_W), \\
&\Sigma_L^{Z\gamma}(k^2) - iM_Z \Sigma^{\gamma\chi}(k^2) - \tilde{\Sigma}^{\gamma Z}(k^2) = -\frac{\alpha}{4\pi}\frac{c_W}{s_W} k^2 B_0(k^2; M_W, M_W), \qquad (3.17)\\
&\Sigma_L^{Z}(k^2) - iM_Z \Sigma^{Z\chi}(k^2) - \tilde{\Sigma}^{Z}(k^2) = \frac{\alpha}{4\pi}\frac{c_W^2}{s_W^2}(k^2 - M_Z^2)\, B_0(k^2; M_W, M_W), \\
&\Sigma_L^{W}(k^2) + M_W \Sigma^{W\varphi}(k^2) - \tilde{\Sigma}^{W}(k^2) \\
&= \frac{\alpha}{4\pi}(k^2 - M_W^2)\left[\frac{c_W^2}{s_W^2} B_0(k^2; M_W, M_Z) + B_0(k^2; M_W, 0)\right],
\end{aligned}$$

where the singular 1-loop integral B_0 is defined in eq. (5.4). Adding the appropriate counter terms of app. A we obtain the identities for the renormalized self energies:

$$\begin{aligned}
&\hat{\Sigma}_L^{\gamma} - \hat{\tilde{\Sigma}}^{\gamma} = k^2\left(\delta Z_2^{\gamma} - \delta\tilde{Z}^{\gamma} - \frac{1}{2}\delta\xi_1^{\gamma} + \frac{\alpha}{4\pi} B_0(k^2; M_W, M_W)\right), \\
&\hat{\Sigma}_L^{\gamma Z} - \hat{\tilde{\Sigma}}^{Z\gamma} = -k^2\left(\delta Z_2^{\gamma Z} - \delta\tilde{Z}^{\gamma Z} - \frac{1}{2}\delta\xi_1^{\gamma Z}\right) + \frac{1}{2} M_Z^2\,(\delta Z_2^{\gamma Z} - 2\delta\tilde{Z}^{\gamma Z}) \\
&\qquad + \frac{\alpha}{4\pi}\frac{c_W}{s_W}(k^2 - M_Z^2)\, B_0(k^2; M_W, M_W), \qquad (3.18)\\
&\hat{\Sigma}_L^{\gamma Z} - iM_Z \hat{\Sigma}^{\gamma\chi} - \hat{\tilde{\Sigma}}^{\gamma Z} = -k^2\left(\delta Z_2^{\gamma Z} - \delta\tilde{Z}^{\gamma Z} - \frac{1}{2}\delta\xi_1^{\gamma Z}\right) \\
&\qquad + \frac{1}{2} M_Z^2(\delta Z_2^{\gamma Z} - \delta\xi_1^{\gamma Z}) - \frac{\alpha}{4\pi}\frac{c_W}{s_W} k^2 B_0(k^2, M_W, M_W), \\
&\hat{\Sigma}_L^{Z} - iM_Z \hat{\Sigma}^{Z\chi} - \hat{\tilde{\Sigma}}^{Z} \\
&= (k^2 - M_Z^2)\left(\delta Z_2^{Z} - \delta\tilde{Z}^{Z} - \frac{1}{2}\delta\xi_1^{Z} + \frac{\alpha}{4\pi}\frac{c_W^2}{s_W^2} B_0(k^2; M_W, M_W)\right), \\
&\hat{\Sigma}_L^{W} + M_W \hat{\Sigma}^{W\varphi} - \hat{\tilde{\Sigma}}^{W} \\
&= (k^2 - M_W^2)\left[\delta Z_2^{W} - \delta\tilde{Z}^{W} - \frac{1}{2}\delta\xi_1^{W} + \frac{\alpha}{4\pi}\left(\frac{c_W^2}{s_W^2} B_0(k^2; M_W, M_Z)\right.\right. \\
&\qquad \left.\left. + B_0(k^2; M_W, 0)\right)\right]. \qquad (3.18')
\end{aligned}$$

3.3 Generalized Ward identities

The identities (3.10) and (3.16) relate unphysical parts of Green functions. In analogy to the QED-Ward identity [24] between the eeγ-vertex and the electron propagator we can derive from eq. (3.8) by differentiating twice with respect to the sources of the fermion fields identities relating fermion vertices to fermion propagators:

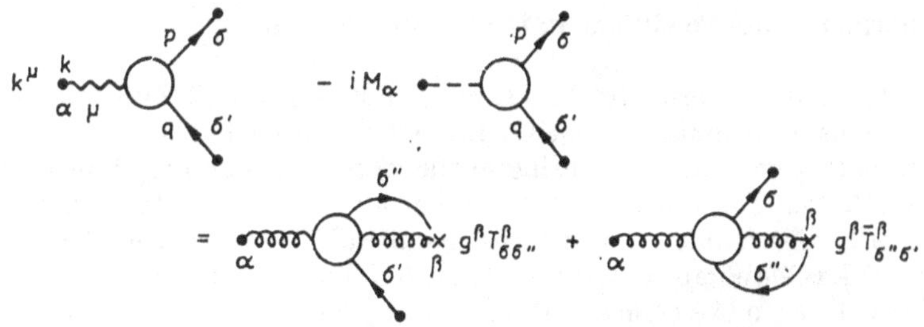

$$k^\mu \tau_\mu{}^{\alpha,\sigma\sigma'}(k, p, q) - iM_\alpha \tau^{\varphi\sigma\sigma'}(k, p, q) = iG^{\alpha\beta}(k) \left[g^\beta T^\beta_{\sigma\sigma''} S_F{}^{\sigma''\sigma'}(q) + S_F{}^{\sigma\sigma''}(p)\, g^\beta \overline{T}^\beta_{\sigma''\sigma'}\right]$$
$$+ ig^\beta T^\beta_{\sigma\sigma''} \int d^4r G^{\alpha\beta\sigma''\sigma'}_{\text{con}}(k, r - p, r, q)$$
$$+ i \int d^4r G^{\alpha\beta,\sigma\sigma''}_{\text{con}}(k, q - r, p, r)\, g^\beta \overline{T}^\beta_{\sigma''\sigma'} \qquad (3.19$$

with

$$\overline{T} = -\gamma_0 T^T \gamma_0.$$

The physical content of these identities can be seen by evaluating them in 1-loop approximation and inserting the results (3.17). Neglecting terms of order $\alpha \cdot m_{i\sigma}/(\sqrt{k^2}, M_W, M_Z)$ the following generalizations of the ordinary QED-Ward identity are valid:

$$k^\mu\left(\Lambda_\mu{}^{\gamma ee}(k, p, q) + \Lambda_\mu{}^{\gamma\nu\nu}(k, p, q)\right) = e[\Sigma^e(p) - \Sigma^e(q)], \qquad (3.20)$$

$$k^\mu\left(\Lambda_\mu{}^{\gamma uu}(k, p, q) - \Lambda_\mu{}^{\gamma dd}(k, p, q)\right) = e\left[\frac{2}{3}\left(\Sigma^u(p) - \Sigma^u(q)\right) - \frac{1}{3}\left(\Sigma^d(p) - \Sigma^d(q)\right)\right]$$

and similar relations for the other fermion generations. The vertex functions $\Lambda_\mu{}^{a\sigma\sigma'}$ are the 1-loop contributions to the amputated 3-point Green functions τ_μ. The fermion self energies $\Sigma^{i\sigma}$ follow from the propagators:

$$iS_F{}^{-1} = \not{p} - m_{i\sigma} + \Sigma^{i\sigma}(p) = \not{p} - m_{i\sigma} + \not{p}\Sigma_V{}^{i\sigma} + \not{p}\gamma_5\Sigma_A{}^{i\sigma} + m_{i\sigma}\Sigma_S{}^{i\sigma}. \qquad (3.21)$$

As a result the QED identities for each separate charged fermion are replaced by similar identities for the fermion doublets. In addition we have found analogous relations for the vertices of the fermions and the heavy gauge boson Z (written for the first fermion doublet):

$$k^\mu(\Lambda_\mu{}^{Z\nu\nu}(k, p, q) + \Lambda_\mu{}^{Zee}(k, p, q)$$
$$= \frac{e}{c_W s_W}\left[\Sigma^\nu(p)\,\frac{1}{2}\cdot\frac{1-\gamma_5}{2} - \frac{1}{2}\cdot\frac{1+\gamma_5}{2}\,\Sigma^\nu(q) + \Sigma^e(p)\left(-\frac{1}{2}\cdot\frac{1-\gamma_5}{2} + s_W{}^2\right)\right.$$
$$\left. + \left(\frac{1}{2}\cdot\frac{1+\gamma_5}{2} - s_W{}^2\right)\Sigma^e(q)\right], \qquad (3.22)$$

$$k^\mu\big(\Lambda_\mu^{Zuu}(k, p, q) - \Lambda_\mu^{Zdd}(k, p, q)\big)$$
$$= \frac{e}{c_W s_W}\left[\Sigma^u(p)\left(\frac{1}{2}\cdot\frac{1-\gamma_5}{2} - \frac{2}{3}\,s_W^2\right) - \left(\frac{1}{2}\cdot\frac{1+\gamma_5}{2} - \frac{2}{3}\,s_W^2\right)\Sigma^u(q)\right.$$
$$\left. + \Sigma^d(p)\left(-\frac{1}{2}\cdot\frac{1-\gamma_5}{2} + \frac{1}{3}\,s_W^2\right) + \left(\frac{1}{2}\cdot\frac{1+\gamma_5}{2} - \frac{1}{3}\,s_W^2\right)\Sigma^d(q)\right].$$

4. Renormalization Conditions in the On-Shell Scheme

The study of the counter terms in the Lagrangian and of the detailed Slavnov-Taylor identities allows us to formulate explicitly the renormalization conditions. Thereby not only the ultraviolet divergencies occuring in the loop expansion are absorbed in the infinite parts of the renormalization constants but also the finite parts are fixed. These lead to physically observable consequences. As already mentioned in the introduction various more or less elaborate renormalizations for the standard model are used in the literature. They differ in the choice of the physical parameters and the prescriptions for the finite parts of the field renormalization constants. Although the results including radiative corrections obtained with different consistent renormalizations formally deviate from each other only on higher order terms it may be that the 1-loop corrections itself calculated with a low energy renormalization scheme and applied to high energy experiments differ from high energy renormalization calculations.

We present a renormalization scheme which is defined by the following conditions[3]):

— The poles of the renormalized propagators lie at M_W^2, M_Z^2, 0, M_H^2, $m_{i\sigma}^2$. This implies for the renormalized self energies:

$$\hat\Sigma_T^W(M_W^2) = \hat\Sigma_T^Z(M_Z^2) = \hat\Sigma^\eta(M_H^2) = \hat\Sigma^{i\sigma}(m_{i\sigma}^2) = 0. \tag{4.1}$$

— According to the residual $U(1)^{\rm em}$ symmetry it is possible to renormalize so that the properties of the photon and the electric charge are defined like in QED:

$$\frac{1}{k^2}\,\hat\Sigma_T^\gamma(k^2)|_{k^2=0} = 0, \quad \hat\Sigma_T^{\gamma Z}(0) = 0, \quad \hat\Gamma_\mu^{\gamma ee}(k^2 = 0, \not p = \not q = m_e) = ie\gamma_\mu{}^{4)}. \tag{4.2}$$

— The residues of the propagators of fermions with $I^3 = -1/2$ and of the physical Higgs particle are one:

$$\left(\frac{1}{\not p - m_{i-}}\,\hat\Sigma^{-i}(p)\right)\bigg|_{\not p = m_{i-}} = 0, \quad \left(\frac{\partial}{\partial p^2}\,\hat\Sigma^\eta(p^2)\right)\bigg|_{p^2 = M_H^2} = 0. \tag{4.3}$$

— Vanishing tadpole:

$$\hat T = 0. \tag{4.4}$$

— The poles in the unphysical sector are at M_W^2, M_Z^2, 0:

$$\hat\Sigma_L^W(M_W^2) = \hat\Sigma_L^Z(M_Z^2) = \hat\Sigma^\varphi(M_W^2) = \hat\Sigma^\chi(M_Z^2) = \hat\Sigma^{\gamma\chi}(0) = 0,$$
$$\frac{1}{k^2}\,\hat\Sigma_L^\gamma|_{k^2=0} = 0. \tag{4.5}$$

[3]) In the following equations only real parts of self energies enter. The imaginary parts are finite by themselves and we define the mass as the real part of the pole position in the propagator.

[4]) This is a condition for the vector part of the photon vertex $\hat\Gamma_\mu{}^{\gamma ee}$, only. For the axial vector part no separate condition has to be imposed since $\hat\Gamma^{ee}_{\mu,A}(k^2 = 0) = 0$ is automatically fulfilled.

— The residue of the photon ghost propagator is one and the photon-ghost Z-ghost mixing propagator vanishes at $k^2 = 0$:

$$\frac{1}{k^2} \hat{\hat{\Sigma}}^{\gamma}(k^2)|_{k^2=0}, \qquad \hat{\hat{\Sigma}}^{Z\gamma}(0) = 0. \tag{4.6}$$

The conditions (4.1)—(4.6) fix all the renormalization constants of eq. (2.17).

We have chosen our scheme in such a way that the following properties hold:

— We use as physical renormalized parameters e, M_W, M_Z, M_H, m_f. The question from which processes an optimal determination of the standard model parameters should be performed depends on the experimental accuracy. At present the best choice is the Josephson effect for the determination of α, the $P\overline{P}$ collider experiments and the μ decay for M_W and M_Z (resp. M_W and $s_W{}^2$), but with experimental progress this may change. Especially a more accurate measurement of the W, Z masses seems to be very desirable, leaving e.g. μ decay and neutrino scattering as low energy tests for the standard model.

— Eqs. (4.2) characterize our procedure as a natural extension of the QED renormalization. This means in practice that existing results on photonic corrections [18, 19] can be taken over directly. Especially 1-loop calculations can be divided into real and virtual photonic corrections (the sum of their contributions in physical cross sections is infrared finite) and weak corrections (IR finite by themselves). We use $e^2/16\pi^2 = \alpha/4\pi = 1/(4\pi \cdot 137.036)$ as the effective expansion parameter.

— We work with only one field renormalization constant for a symmetry multiplet. Therefore renormalization conserves the gauge transformation properties of the fields and the Green functions. But as a consequence of the use of the minimal number of field renormalization constants not all the residues of the renormalized propagators are one. This is the case for the W, Z and the $I^3 = +1/2$ fermions and wave function renormalization for their in- and outgoing particles is needed. These do not occur in the complete amplitudes for physical S matrix elements.

— In $\mathcal{L}_{\text{fix}} + \mathcal{L}_{FP}$ we have built in the renormalization constants $\delta\xi^{a}_{1,2}$ and fixed them in such a way that the simple pole structure of the 't Hooft-Feynman gauge survives renormalization. The Slavnov-Taylor identities (3.15) and (3.18) guarantee that with the conditions (4.5) also the poles in the other unphysical propagators $\Delta^{W\varphi}; \ldots; G^W; \ldots$ are at the same positions. This simplifies considerably the evaluation of Feynman diagrams.

— We have checked that the Ward identities for the fermion gauge boson vertices (3.19′) are compatible with our renormalization prescription.

Finally we translate the conditions (4.1)—(4.6) into prescriptions for the singular and finite parts of the renormalization constants (2.17) resp. their combinations (app. A):

$$\Sigma_T{}^W(M_W{}^2) = \delta\overline{M}_W{}^2 = \delta M_W{}^2 + 2M_W{}^2 \frac{\delta t}{t}$$
$$= M_W{}^2 \left(-2\frac{\delta v}{v} + 2\delta Z_1{}^W - 3\delta Z_2{}^W + \delta Z^\varphi + 2\frac{\delta t}{t}\right),$$

$$\Sigma_T{}^Z(M_Z{}^2) = \delta\overline{M}_Z{}^2 = \delta M_Z{}^2 + 2M_Z{}^2 \frac{\delta t}{t}$$
$$= M_Z{}^2 \left(-2\frac{\delta v}{v} + 2\delta Z_1{}^Z - 3\delta Z_2{}^Z + \delta Z^\varphi + 2\frac{\delta t}{t}\right), \tag{4.1'}$$

$$\Sigma^{\eta}(M_H^2) = \delta \overline{M}_H{}^2 = \delta M_H{}^2 + 3M_H{}^2 \frac{\delta t}{t}$$

$$= M_H{}^2 \left(-3 \frac{\delta v}{v} + \frac{3}{2} \delta Z^{\lambda} - \delta Z^{\varphi} + 3 \frac{\delta t}{t}\right) + \delta \mu^2, \tag{4.1'}$$

$$\Sigma_V{}^{i\sigma}(m_{i\sigma}) + \Sigma_S{}^{i\sigma}(m_{i\sigma}) = \frac{\delta \overline{m}_{i\sigma}}{m_{i\sigma}} = \frac{\delta m_{i\sigma}}{m_{i\sigma}} + \frac{\delta t}{t} = \delta Z_1{}^{i\sigma} - \frac{\delta v}{v} + \frac{\delta t}{t};$$

$$\frac{1}{k^2} \Sigma_T{}^{\gamma}(k^2)|_{k^2=0} = -\delta Z_2{}^{\gamma},$$

$$\Sigma_\tau{}^{\gamma Z}(0) = -M_Z{}^2(\delta Z_1{}^{\gamma Z} - \delta Z_2{}^{\gamma Z}), \tag{4.2'}$$

$$\frac{\delta e}{e} = \delta Z_1{}^{\gamma} - \frac{3}{2} \delta Z_2{}^{\gamma};$$

$$\Sigma_L{}^e(m_e) + m_e{}^2\big(\Sigma_L{}^{e'}(m_e) + \Sigma_R{}^{e'}(m_e) + 2\Sigma_s{}^{e'}(m_e)\big) = -\delta Z_L{}^{(e,\nu_e)},$$

$$\Sigma_R{}^e(m_e) + m_e{}^2\big(\Sigma_L{}^{e'}(m_e) + \Sigma_R{}^{e'}(m_e) + 2\Sigma_s{}^{e'}(m_e)\big) = -\delta Z_R{}^e, \tag{4.3'}$$

$$\Sigma\,'(M_H{}^2) = -\delta Z^{\varphi};$$

$$\frac{\delta t}{t} = \frac{\delta v}{v} - \frac{\delta \mu^2}{M_H{}^2} - \frac{1}{2} \delta Z^{\lambda}, \qquad t = M_W M_H{}^2 \cdot \frac{2 s_W}{e}; \tag{4.4'}$$

$$\Sigma_L{}^W(M_W{}^2) = \delta \overline{M}_W{}^2 + M_W{}^2 \delta \xi_1{}^W,$$

$$\Sigma_L{}^Z(M_Z{}^2) = \delta \overline{M}_Z{}^2 + M_Z{}^2 \delta \xi_1{}^Z,$$

$$\Sigma^{\varphi}(M_W{}^2) = \delta M_W{}^2 + M_W{}^2 \delta \xi_2{}^W,$$

$$\Sigma^{\chi}(M_Z{}^2) = \delta M_Z{}^2 + M_Z{}^2 \delta \xi_2{}^Z,$$

$$\frac{1}{k^2} \Sigma_L{}^{\gamma}(k^2)|_{k^2=0} = \delta \xi_1{}^{\gamma} - \delta Z_2{}^{\gamma}, \tag{4.5'}$$

$$\Sigma^{\gamma\chi}(0) = -\frac{1}{2} M_Z(\delta \xi_1{}^{\gamma Z} - \delta \xi_2{}^{\gamma Z}),$$

$$\frac{1}{k^2} \tilde{\Sigma}^{\gamma}(k^2)|_{k^2=0} = -\delta \tilde{Z}_2{}^{\gamma} + \frac{1}{2} \delta \xi_1{}^{\gamma}, \tag{4.6'}$$

$$\tilde{\Sigma}^{Z\gamma}(0) = M_Z{}^2 \left(\delta Z_1{}^{\gamma Z} - \frac{3}{2} \delta Z_2{}^{\gamma Z} + \delta \tilde{Z}^{\gamma Z}\right).$$

5. Explicit Results in 1-Loop Approximation

The intention of this section is to provide the building blocks needed to compute radiative electroweak corrections to e^+e^- annihilation, deep inelastic scattering and other processes. We do this by evaluating the explicit 1-loop expressions for the renormalization constants, renormalized self energies and fermion gauge boson vertices. The calculations are performed analytically, thereby neglecting terms which are of the order of magnitude $\alpha m_f^2/(M_W^2, s)$ in the final results. The ultraviolet divergences are treated with the method of dimensional regularization [25]. This is possible since the standard model is free of γ_5-anomalies. The 4-dimensional integration and the Dirac and tensor structures are replaced by D-dimensional ones:

$$\int \frac{d^4q}{(2\pi)^4} \to \mu^{4-D} \int \frac{d^Dq}{(2\pi)^D} \tag{5.1}$$

(μ is introduced for dimensional reasons).

In order to explain our notation we give the results for the scalar tadpole integral:

q

$$\to \mu^{4-D} \int \frac{d^Dq}{(2\pi)^D} \frac{1}{q^2 - M^2 + i\varepsilon} = \frac{-i}{16\pi^2} A(M)$$
$$A(M) = -M^2(\Delta_M + 1). \tag{5.2}$$

The UV divergent part Δ_M contains the Euler-Mascheroni constant γ and has the form:

$$\Delta_M = \frac{2}{4-D} - \gamma - \ln \frac{M^2}{4\pi\mu^2}. \tag{5.3}$$

The scalar 1-loop self energy integral defines the function $B_0(k^2; M_1, M_2)$:

q + k

q

$$\to \mu^{4-D} \int \frac{d^Dq}{(2\pi)^D} \frac{1}{(q^2 - M_1^2 + i\varepsilon)\left((q+k)^2 - M_2^2 + i\varepsilon\right)}$$
$$= \frac{i}{16\pi^2} B_0(k^2; M_1, M_2)$$

with:

$$B_0(k^2; M_1, M_2) = \frac{1}{2}\Delta_{M_1} + \frac{1}{2}\Delta_{M_2} + 1 - \frac{M_1^2 + M_2^2}{M_1^2 - M_2^2} \ln \frac{M_1}{M_2} + F(k^2; M_1, M_2)$$
$$= \frac{1}{2}\Delta_{M_1} + \frac{1}{2}\Delta_{M_2} - \int_0^1 dx \ln \frac{x^2k^2 - x(k^2 + M_1^2 - M_2^2) + M_1^2 - i\varepsilon}{M_1 M_2}. \tag{5.4}$$

The function $F(k^2; M_1, M_2) = F(k^2; M_2, M_1)$ is finite and vanishes for $k^2 = 0$. Its explicit form is written in app. B.1.

5.1 The tadpole

The vacuum expectation value v of the Higgs field, which in lowest order is given by $v^2 = \mu^2/\lambda$ gets 1-loop contributions from the diagrams of fig. 1. They lead in the 't Hooft-

Feynman gauge to the expressions:

$$T = \frac{1}{16\pi^2}\frac{e}{s_W}\cdot\frac{1}{2M_W}\left\{-M_H{}^2\left(A(M_W)+\frac{1}{2}A(M_Z)+\frac{3}{2}A(M_H)\right)\right.$$
$$\left.-6M_W{}^2A(M_W)-4M_W{}^4-3M_Z{}^2A(M_Z)-2M_Z{}^4+4\sum_{i\sigma}m_{i\sigma}^2A(m_{i\sigma})\right\}$$
$$= -\delta t = -\frac{2s_W}{e}M_WM_H{}^2\frac{\delta t}{t}. \tag{5.5}$$

The tadpole diagrams of fig. 1 give for example contributions to the self energies. These are absorbed by mass renormalization rending the $\delta\overline{M}_i{}^2$ gauge independent.

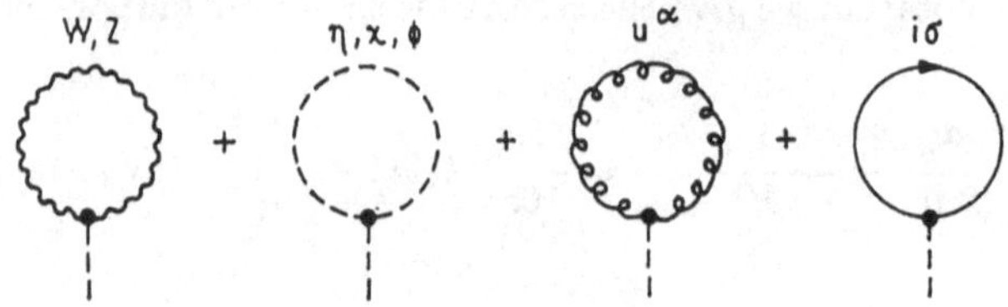

Fig. 1. 1-loop tadpole diagrams

5.2 Unrenormalized self energies and vertex functions

a) Gauge boson self energies

The contributions of the diagrams of fig. 2 to the longitudinal and transverse unrenormalized self energies have been computed by [9]. We present them decomposed into the singular parts (defined to be proportional to Δ) and finite parts:

$$\Sigma(k^2) = \Sigma_{\text{sing}}(k^2, \Delta) + \Sigma_{\text{fin}}(k^2). \tag{5.6}$$

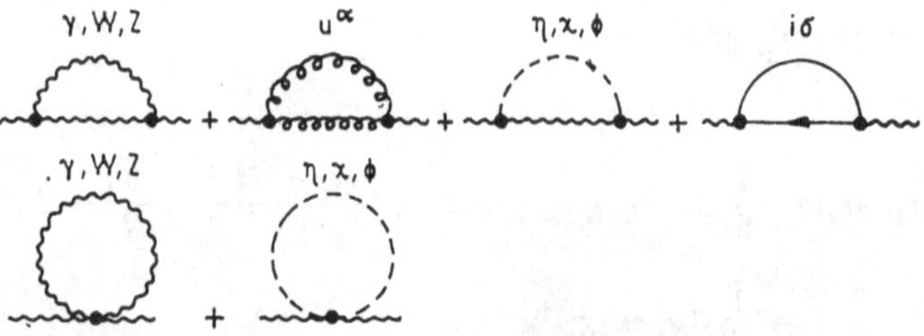

Fig. 2. 1-loop gauge boson self energy diagrams*).

The explicit expressions are (the index f denotes any fermion $i\sigma$; $v_f = v_{i\sigma}$, $a_f = a_{i\sigma}$ are defined in app. A. We also omit in sects. 5 and 6 the index W at s_W, c_W):

$$\Sigma^{\gamma}_{T,\text{sing}} = \frac{\alpha}{4\pi}k^2\left\{\frac{4}{3}\sum_f Q_f{}^2\Delta_f - 3\Delta_W\right\},$$
$$\Sigma^{\gamma}_{T,\text{fin}} = \frac{\alpha}{4\pi}\left\{\frac{4}{3}\sum_f Q_f{}^2\left[(k^2+2m_f{}^2)\,F(k^2; m_f, m_f) - \frac{k^2}{3}\right]\right.$$
$$\left. - (3k^2 + 4M_W{}^2)\,F(k^2; M_W, M_W)\right\}, \tag{5.7}$$

*) In figs. 2—7 tadpole diagrams are omitted.

$$\Sigma^{\gamma Z}_{T,\text{sing}} = \frac{\alpha}{4\pi}\left\{-\frac{4}{3}\sum_f Q_f v_f k^2 \Delta_f + \frac{1}{cs}\left[k^2\left(3c^2+\frac{1}{6}\right)+2M_W^2\right]\Delta_W\right\},$$

$$\Sigma^{\gamma Z}_{T,\text{fin}} = \frac{\alpha}{4\pi}\left\{-\frac{4}{3}\sum_f Q_f v_f\left[(k^2+2m_f^2)\,F(k^2;m_f,m_f)-\frac{k^2}{3}\right]\right.$$
$$\left.+\frac{1}{cs}\left[k^2\left(3c^2+\frac{1}{6}\right)+M_W^2\left(4c^2+\frac{4}{3}\right)\right]F(k^2;M_W,M_W)+\frac{k^2}{9cs}\right\}; \tag{5.8}$$

$$\Sigma^{Z}_{T,\text{sing}} = \frac{\alpha}{4\pi}\left\{\frac{4}{3}\sum_f\left[(v_f^2+a_f^2)\,k^2-\frac{3m_f^2}{8c^2s^2}\right]\Delta_{\tilde f}\right.$$
$$\left.+\left[k^2\left(3-\frac{19}{6s^2}+\frac{1}{6c^2}\right)+M_Z^2\left(4+\frac{1}{c^2}-\frac{1}{s^2}\right)\right]\Delta_W\right\}-2M_Z^2\,\frac{T}{t},^{5)} \tag{5.9}$$

$$\Sigma^{Z}_{T,\text{fin}} = \frac{\alpha}{4\pi}\left\{\frac{4}{3}\sum_f\left[(v_f^2+a_f^2)\left((k^2+2m_f^2)\,F(k^2;m_{\tilde f},m_{\tilde f})-\frac{k^2}{3}\right)\right.\right.$$
$$\left.-\frac{3}{8c^2s^2}\,m_f^2F(k^2;m_f,m_f)\right]$$
$$-\frac{c^2}{3s^2}\left[\frac{2}{3}k^2+(10k^2+20M_W^2)\,F(k^2;M_W,M_W)\right]$$
$$+\frac{1}{3s^2c^2}\left[3M_W^2F(k^2;M_W,M_W)+\frac{1}{4}(10M_Z^2-2M_H^2+k^2)\right.$$
$$\times\left(1-\frac{M_H^2+M_Z^2}{M_H^2-M_Z^2}\ln\frac{M_H}{M_Z}-\ln\frac{M_HM_Z}{M_W^2}+F(k^2;M_H,M_Z)\right)$$
$$-\frac{1}{2}M_H^2\ln\frac{M_H^2}{M_W^2}-\frac{1}{2}M_Z^2\ln\frac{M_Z^2}{M_W^2}+\frac{k^2}{6}$$
$$\left.+(M_H^2-M_Z^2)^2\,F(k^2;M_Z,M_H)/4k^2\right]+\frac{(c^2-s^2)^2}{3s^2c^2}$$
$$\left.\times\left[\frac{k^2}{6}+\left(2M_W^2+\frac{k^2}{4}\right)F(k^2;M_W,M_W)\right]\right\};$$

$$\Sigma^{W}_{T,\text{sing}} = \frac{\alpha}{4\pi}\cdot\frac{1}{s^2}\left\{\frac{1}{6}\sum_i\left[\Delta_{i+}\left(k^2-\frac{5}{2}m_{i+}^2-\frac{1}{2}m_{i-}^2\right)\right.\right.$$
$$\left.\left.+\Delta_{i-}\left(k^2-\frac{5}{2}m_{i-}^2-\frac{1}{2}m_{i+}^2\right)\right]-\left[M_W^2\left(1-\frac{s^2}{c^2}\right)+\frac{19}{6}k^2\right]\Delta_W\right\}$$
$$-2M_W^2T/t,^{6)} \tag{5.10}$$

$$\Sigma^{W}_{T,\text{fin}} = \frac{\alpha}{4\pi}\left\{\frac{1}{3s^2}\sum_i\left[\left(k^2-\frac{m_{i+}^2+m_{i-}^2}{2}\right)\left(1-\frac{m_{i+}^2+m_{i-}^2}{m_{i+}^2-m_{i-}^2}\ln\frac{m_{i+}}{m_{i-}}\right.\right.\right.$$
$$\left.\left.+F(k^2;m_{i+},m_{i-})\right)-\frac{k^2}{3}-\frac{(m_{i+}^2-m_{i-}^2)^2}{2k^2}F(k^2;m_{i+},m_{i-})\right]$$

[5]) $\tilde f = f$ for $f \neq \nu_l$ and $\tilde f = l$ for $f = \nu_l$.

[6]) For lepton doublets with $m_{i+} = 0$ replace $\Delta_{i+} \to \Delta_{i-}$ and drop the log term in $\Sigma^W_{T,\text{fin}}$.

$$- \frac{c^2}{3s^2} \left[(7M_Z^2 + 7M_W^2 + 10k^2) \left(1 - \frac{M_Z^2}{M_Z^2 - M_W^2} \ln \frac{M_Z^2}{M_W^2} \right.\right.$$

$$\left. + F(k^2; M_Z, M_W) \right) + 4M_Z^2 \ln \frac{M_Z^2}{M_W^2} + \frac{2}{3} k^2 - \frac{2}{k^2} (M_Z^2 - M_W^2)^2$$

$$\left. \times F(k^2; M_Z, M_W) \right] - \frac{4}{3} M_W^2 - \frac{32}{9} k^2 - \frac{1}{3} (4M_W^2 + 10k^2)$$

$$\times F(k^2; 0, M_W) + \frac{2}{3} \frac{M_W^4}{k^2} F(k^2; 0, M_W)$$

$$+ \frac{s^2}{c^2} \left[1 - \frac{M_Z^2}{M_Z^2 - M_W^2} \ln \frac{M_Z^2}{M_W^2} + F(k^2; M_Z, M_W) \right] M_W^2$$

$$+ \frac{1}{s^2} \left[1 - \frac{M_H^2}{M_H^2 - M_W^2} \ln \frac{M_H^2}{M_W^2} + F(k^2; M_H, M_W) \right] M_W^2$$

$$+ \frac{1}{s^2} \left[\frac{5}{18} k^2 - \frac{1}{3} M_W^2 - \frac{1}{6} M_Z^2 - \frac{1}{6} M_H^2 \right.$$

$$+ \frac{1}{6} \left(2M_W^2 - \frac{1}{2} k^2 \right) \frac{M_Z^2}{M_Z^2 - M_W^2} \ln \frac{M_Z^2}{M_W^2}$$

$$- \frac{1}{6} \left(M_W^2 + M_Z^2 - \frac{k^2}{2} \right) F(k^2; M_Z, M_W)$$

$$+ (M_Z^2 - M_W^2)^2 F(k^2; M_Z, M_W)/12k^2$$

$$+ \frac{1}{6} \left(2M_W^2 - \frac{1}{2} k^2 \right) \frac{M_H^2}{M_H^2 - M_W^2} \ln \frac{M_H^2}{M_W^2}$$

$$- \frac{1}{6} \left(M_W^2 + M_H^2 - \frac{k^2}{2} \right) F(k^2; M_H, M_W)$$

$$\left.\left. + (M_H^2 - M_W^2)^2 F(k^2; M_H, M_W)/12k^2 \right] \right\};$$

$$\Sigma^{\gamma}_{L,\mathrm{sing}} = \Sigma^{\gamma}_{L,\mathrm{fin}} = 0; \tag{5.11}$$

$$\Sigma^{Z}_{L,\mathrm{sing}} = -\frac{\alpha}{4\pi} \left\{ M_Z^2 \left(2\frac{c^2}{s^2} - \frac{1}{c^2 s^2} - 2 \right) \Delta_W \right\} - 2M_Z^2 T/t,$$

$$\Sigma^{Z}_{L,\mathrm{fin}} = \frac{\alpha}{4\pi} \left\{ \frac{1}{c^2 s^2} \left[M_Z^2 - \frac{M_Z^2}{M_H^2 - M_Z^2} \left(M_H^2 \ln \frac{M_H^2}{M_W^2} - M_Z^2 \ln \frac{M_Z^2}{M_W^2} \right) \right.\right.$$

$$\left. + \left(M_Z^2 - \frac{(M_H^2 - M_Z^2)^2}{4k^2} \right) F(k^2; M_H, M_Z) \right]$$

$$\left. - 2 \frac{c^2 - s^2}{s^2} M_Z^2 F(k^2; M_W, M_W) \right\}; \tag{5.12}$$

$$\Sigma^{\gamma Z}_{L,\mathrm{sing}} = \frac{\alpha}{4\pi} \left\{ 2 \frac{c}{s} M_Z^2 \Delta_W \right\},$$

$$\Sigma^{\gamma Z}_{L,\mathrm{fin}} = \frac{\alpha}{4\pi} \left\{ 2 \frac{c}{s} M_Z^2 F(k^2; M_W, M_W) \right\}; \tag{5.13}$$

$$\Sigma^W_{L,\text{sing}} = -\frac{\alpha}{4\pi}\left\{\frac{c^2 - s^2}{c^2 s^2} M_W{}^2 \Delta_W\right\} - 2M_W{}^2 T/t,$$

$$\begin{aligned}\Sigma^W_{L,\text{fin}} = -\frac{\alpha}{4\pi}\Bigg\{&\frac{c^2 - s^2}{c^2 s^2} M_W{}^2 + \frac{3s^2 - 2}{s^4} M_Z{}^2 \ln\frac{M_Z{}^2}{M_W{}^2} \\ &+ \frac{1}{s^2}\frac{M_W{}^2}{M_H{}^2 - M_W{}^2} M_H{}^2 \ln\frac{M_H{}^2}{M_W{}^2} + 2\frac{M_W{}^4}{k^2} F(k^2; 0, M_W) \\ &+ \left[\frac{3c^2 - 1}{c^2 s^2} M_W{}^2 + \left(2\frac{c^2}{s^2} + \frac{1}{4s^2}\right)\frac{(M_W{}^2 - M_Z{}^2)^2}{k^2}\right] F(k^2; M_W, M_Z) \\ &- \frac{1}{s^2}\left(M_W{}^2 - \frac{(M_H{}^2 - M_W{}^2)^2}{4k^2}\right) F(k^2; M_W, M_H)\Bigg\}.\end{aligned} \tag{5.14}$$

b) Gauge boson Higgs boson mixing

The diagrams of fig. 3 contribute to the gauge boson Higgs boson mixing energies defined in eq. (3.12). Their singular and finite parts are

Fig. 3. Diagrams for gauge boson Higgs boson mixing

$$i\Sigma^{\gamma\chi}_{\text{sing}} = \frac{\alpha}{4\pi}\left\{2\frac{c}{s} M_Z \Delta_W\right\},$$

$$i\Sigma^{\gamma\chi}_{\text{fin}} = \frac{\alpha}{4\pi}\left\{2\frac{c}{s} M_Z F(k^2; M_W, M_W)\right\}; \tag{5.15}$$

$$i\Sigma^{Z\chi}_{\text{sing}} = \frac{\alpha}{4\pi}\left\{2 + \frac{1}{2c^2} + \frac{1}{4c^2 s^2}\right\} M_Z \Delta_W - M_Z T/t,$$

$$\begin{aligned}i\Sigma^{Z\chi}_{\text{fin}} = \frac{\alpha}{4\pi} M_Z \Bigg\{&\frac{3}{4c^2 s^2}\left[1 - \frac{1}{M_H{}^2 - M_Z{}^2}\left(M_H{}^2 \ln\frac{M_H{}^2}{M_W{}^2} - M_Z{}^2 \ln\frac{M_Z{}^2}{M_W{}^2}\right)\right. \\ &\left.+ \left(1 - \frac{(M_Z{}^2 - M_H{}^2)^2}{3k^2 M_Z{}^2}\right) F(k^2; M_H, M_Z)\right] + \frac{4s^2 - 1}{2s^2} F(k^2; M_W, M_W)\Bigg\};\end{aligned} \tag{5.16}$$

$$\Sigma^{W\varphi}_{\text{sing}} = \frac{\alpha}{4\pi} M_W \left\{\frac{1}{2s^2} - \frac{3}{4c^2 s^2}\right\} \Delta_W + M_W T/t,$$

$$\begin{aligned}\Sigma^{W\varphi}_{\text{fin}} = \frac{\alpha}{4\pi} M_W \Bigg\{&-\frac{1}{4s^2} - \frac{3}{4c^2 s^2} \\ &+ \frac{3}{4s^2}\left[1 - \frac{M_H{}^2}{M_W{}^2 - M_H{}^2} \ln\frac{M_H{}^2}{M_W{}^2}\right. \\ &\left.- \left(1 - \frac{(M_H{}^2 - M_W{}^2)^2}{3k^2 M_W{}^2}\right) F(k^2; M_W, M_H)\right]\end{aligned} \tag{5.17}$$

$$+ \frac{4c^2s^2 + 5c^2 - 3}{4c^2s^2} \left[\frac{M_Z^2}{M_W^2 - M_Z^2} \ln \frac{M_Z^2}{M_W^2} + F(k^2; M_W, M_Z)\right]$$
$$- \frac{8c^2 + 1}{4c^2} \frac{M_W^2 - M_Z^2}{k^2} F(k^2; M_W, M_Z) + \left(2\frac{M_W^2}{k^2} - 1\right) F(k^2; 0, M_W)\bigg\}.$$

c) Higgs boson self energies

The self energies of the Higgs bosons are needed to calculate the full set of renormalization constants. The diagrams of fig. 4 give

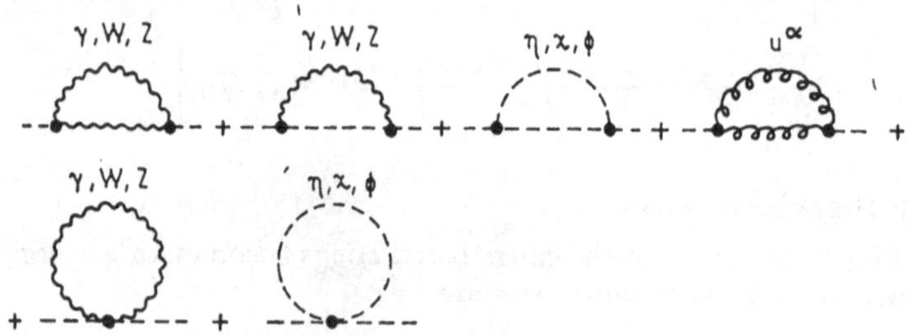

Fig. 4. 1-loop Higgs boson self energy diagrams

$$\Sigma^\eta_{\rm sing} = -\frac{\alpha}{4\pi} \left\{\frac{2c^2+1}{2c^2s^2} k^2 - \frac{17}{4c^2s^2} M_Z^2 - \frac{17}{2s^2} M_W^2 - \frac{2c^2+1}{8c^2s^2} M_H^2 - \frac{27}{8s^2} \frac{M_H^4}{M_W^2}\right\} \Delta_W - 3M_H^2 T/t,$$

$$\Sigma^\eta_{\rm fin} = -\frac{\alpha}{4\pi} \left\{\frac{7}{4c^2s^2} (M_Z^2 + 2c^2M_W^2) - \frac{2c^2+1}{8c^2s^2} M_H^2 - \frac{3}{8s^2} \frac{M_H^4}{M_W^2} \right.$$
$$+ \frac{1}{4c^2s^2} \left[-2k^2 + 17M_Z^2 + \frac{1}{2} M_H^2 + \frac{M_H^4}{M_Z^2}\right] \ln \frac{M_Z^2}{M_W^2}$$
$$+ \frac{21}{8s^2} \frac{M_H^4}{M_W^2} \ln \frac{M_H^2}{M_W^2} - \frac{9}{4s^2} \frac{M_H^4}{M_W^2} F(k^2; M_H, M_H)$$
$$+ \frac{1}{s^2} \left[k^2 - 7M_W^2 - \frac{M_H^4}{2M_W^2}\right] F(k^2; M_W, M_W)$$
$$\left. + \frac{1}{2c^2s^2} \left[k^2 - 7M_Z^2 - \frac{M_H^4}{2M_W^2}\right] F(k^2; M_Z, M_Z)\right\}; \tag{5.18}$$

$$\Sigma^\chi_{\rm sing} = -\frac{\alpha}{4\pi} \frac{2c^2+1}{2c^2s^2} k^2 \Delta_W,$$

$$\Sigma^\chi_{\rm fin} = -\frac{\alpha}{4\pi} \left\{\frac{1}{2c^2s^2} k^2 \left[1 - \frac{1}{M_H^2 - M_Z^2} \left(M_H^2 \ln \frac{M_H^2}{M_W^2} - M_Z^2 \ln \frac{M_Z^2}{M_W^2}\right) \right.\right.$$
$$\left.\left. + \left(1 - \frac{(M_H^2 - M_Z^2)^2}{2k^2M_Z^2}\right) F(k^2; M_H, M_Z)\right] + \frac{1}{s^2} k^2 F(k^2; M_W, M_W)\right\}; \tag{5.19}$$

$$\Sigma^\varphi_{\rm sing} = -\frac{\alpha}{4\pi} \frac{2c^2+1}{2c^2s^2} k^2 \Delta_W, \tag{5.20}$$

$$\Sigma^{\varphi}_{\text{fin}} = -\frac{\alpha}{4\pi}\left\{\frac{2c^2+1}{2c^2s^2}k^2 + \frac{1+c^2-4c^2s^2}{2c^2s^2}k^2 \right.$$
$$\times \left[\frac{M_Z^2}{M_W^2-M_Z^2}\ln\frac{M_Z^2}{M_W^2} + F(k^2; M_W, M_Z)\right]$$
$$+ \frac{k^2M_H^2}{2s^2(M_W^2-M_H^2)}\ln\frac{M_H^2}{M_W^2} - \frac{(8c^2+1)\,s^2}{4c^2}M_Z^2F(k^2; M_W, M_Z)$$
$$+ \frac{1}{4s^2}\left(2k^2 - \frac{(M_W^2-M_H^2)^2}{M_W^2}\right)F(k^2; M_W, M_H)$$
$$\left. + 2(k^2 - M_W^2)\,F(k^2; 0, M_W)\right\}.$$

d) Ghost self energies

The results of the diagrams of fig. 5 for the ghost self energies are:

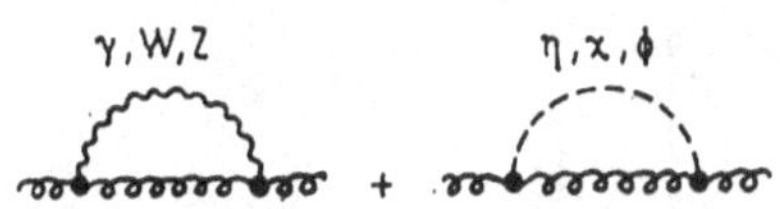

Fig. 5. 1-loop ghost self energy diagrams

$$\tilde{\Sigma}^{\gamma}_{\text{sing}} = -\frac{\alpha}{4\pi}k^2\Delta_W, \qquad \tilde{\Sigma}^{\gamma}_{\text{fin}} = -\frac{\alpha}{4\pi}k^2F(k^2; M_W, M_W); \tag{5.21}$$

$$\tilde{\Sigma}^{Z\gamma}_{\text{sing}} = \frac{\alpha}{4\pi}\left\{\frac{c}{s}(k^2+M_Z^2)\,\Delta_W\right\}, \tag{5.22}$$

$$\tilde{\Sigma}^{Z\gamma}_{\text{fin}} = \frac{\alpha}{4\pi}\left\{\frac{c}{s}(k^2+M_Z^2)\,F(k^2; M_W, M_W)\right\};$$
$$\tilde{\Sigma}^{\gamma Z}_{\text{sing}} = \frac{\alpha}{4\pi}\left\{\frac{c}{s}k^2\Delta_W\right\}, \qquad \tilde{\Sigma}^{\gamma Z}_{\text{fin}} = \frac{\alpha}{4\pi}\left\{\frac{c}{s}k^2F(k^2; M_W, M_W)\right\}; \tag{5.23}$$

$$\tilde{\Sigma}^{Z}_{\text{sing}} = -\frac{\alpha}{4\pi}\left\{\frac{c^2}{s^2}k^2 - \left(1+\frac{1}{4c^2s^2}-\frac{1}{2s^2}\right)M_Z^2\right\}\Delta_W - M_Z^2T/t,$$
$$\tilde{\Sigma}^{Z}_{\text{fin}} = -\frac{\alpha}{4\pi}\left\{-\frac{M_Z^2}{4c^2s^2}\left[1-\frac{1}{M_H^2-M_Z^2}\left(M_H^2\ln\frac{M_H^2}{M_W^2} - M_Z^2\ln\frac{M_Z^2}{M_W^2}\right)\right.\right.$$
$$\left.\left. + F(k^2; M_H, M_Z)\right] + \frac{c^2}{s^2}\left(k^2 - \frac{2s^2-1}{2c^2}M_Z^2\right)F(k^2; M_W, M_W)\right\}; \tag{5.24}$$

$$\tilde{\Sigma}^{W}_{\text{sing}} = -\frac{\alpha}{4\pi}\left\{\frac{1}{s^2}k^2 + \frac{2c^2-1}{4c^2s^2}M_W^2\right\}\Delta_W - M_W^2T/t,$$
$$\tilde{\Sigma}^{W}_{\text{fin}} = -\frac{\alpha}{4\pi}\left\{\frac{1}{s^2}k^2 + \frac{2c^2-1}{4c^2s^2}M_W^2 - \frac{1}{s^2}\left(\frac{c^2}{s^2}k^2 + \frac{3c^2-1}{4c^2s^2}M_W^2\right)\ln\frac{M_Z^2}{M_W^2}\right.$$
$$+ \frac{1}{4s^2}\frac{M_H^2M_W^2}{M_H^2-M_W^2}\ln\frac{M_H^2}{M_W^2} + k^2F(k^2; M_W, 0)$$
$$\left. + \left(\frac{c^2}{s^2}k^2 + \frac{3c^2-1}{4c^2s^2}M_W^2\right)F(k^2; M_W, M_Z) - \frac{1}{4s^2}M_W^2F(k^2; M_W, M_H)\right\}. \tag{5.25}$$

e) Fermion self energies

Because of Lorentz covariance we can decompose the self energies $\Sigma^{i\sigma}(k)$ of the fermions:

$$\Sigma^{i\sigma}(k) = \not{k}\Sigma_V^{i\sigma}(k^2) + \not{k}\gamma_5\Sigma_A^{i\sigma}(k^2) + m_{i\sigma}\Sigma_S^{i\sigma}(k^2). \tag{5.26}$$

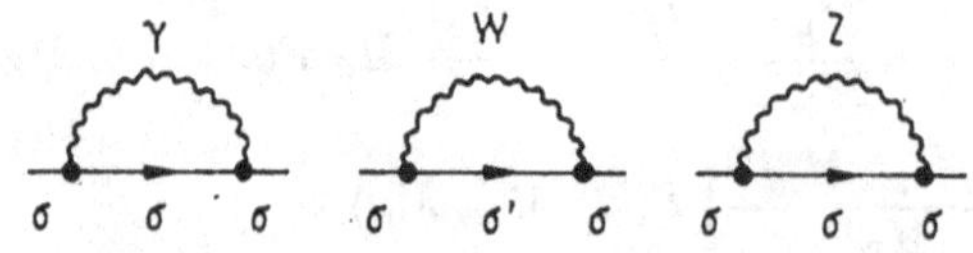

Fig. 6. 1-loop fermion self energy diagrams.

The diagrams of fig. 6 give the following contributions to the invariant functions $\Sigma^{i\sigma}_{V,A,S}$:

$$\begin{aligned}\Sigma_V^{i\sigma} &= -\frac{\alpha}{4\pi}\Big[Q_{i\sigma}^2\big(2B_1(k^2; m_{i\sigma}, \lambda) + 1\big) + (v_{i\sigma}^2 + a_{i\sigma}^2)\big(2B_1(k^2; m_{i\sigma}, M_Z) + 1\big) \\ &\quad + \frac{1}{4s^2}\big(2B_1(k^2; m_{i\sigma}, M_W) + 1\big)\Big], \\ \Sigma_A^{i\sigma} &= -\frac{\alpha}{4\pi}\Big[2v_{i\sigma}a_{i\sigma}\big(2B_1(k^2; m_{i\sigma}, M_Z) + 1\big) - \frac{1}{4s^2}\big(2B_1(k^2; m_{i\sigma'}, M_W) + 1\big)\Big], \\ \Sigma_S^{i\sigma} &= -\frac{\alpha}{4\pi}\big[Q_{i\sigma}^2\big(4B_0(k^2; m_{i\sigma}, \lambda) - 2\big) + (v_{i\sigma}^2 - a_{i\sigma}^2)\big(4B_0(k^2; m_{i\sigma}, M_Z) - 2\big)\big].\end{aligned} \tag{5.27}$$

The photon contribution was calculated with a small photon mass λ in order to regularize possible infrared divergencies. The functions B_0 and B_1 are defined in eqs. (5.4) and (B.2). Instead of the vector and axial vector parts of the self energies $\Sigma_{V,A}$ it may be more convenient to use the right- and left-handed parts:

$$\Sigma_R = (\Sigma_V + \Sigma_A), \qquad \Sigma_L = (\Sigma_V - \Sigma_A). \tag{5.27'}$$

f) Fermion gauge boson vertex functions

The vertex functions[7]) $\Gamma_\mu^{a\sigma\sigma'}(k^2, p, q)$ contain for $|k^2| \gg m_{\sigma\sigma'}^2$ and $p^2 = m_\sigma^2$, $q^2 = m_\sigma^2$, only vector and axial vector parts. The Feynman diagrams of fig. 7 yield (diagrams containing Higgs exchanges can be neglected):

$$\begin{aligned}\Gamma_\mu^{\gamma\sigma\sigma}(k^2) &= -ieQ_\sigma\gamma_\mu \\ &\quad - ieQ_\sigma\gamma_\mu\frac{\alpha}{4\pi}Q_\sigma^2\left[\Delta_\sigma - 2\ln\frac{m_\sigma^2}{\lambda^2} + 4 + \Lambda_1(k^2, m_\sigma)\right] \\ &\quad - ieQ_\sigma[(v_\sigma^2 + a_\sigma^2)\gamma_\mu - 2v_\sigma a_\sigma\gamma_\mu\gamma_5]\frac{\alpha}{4\pi}\left[\Delta_Z - \frac{1}{2} + \Lambda_2(k^2, M_Z)\right] \\ &\quad - ieQ_{\sigma'}\gamma_\mu(1-\gamma_5)\frac{\alpha}{4\pi}\frac{1}{4s^2}\left[\Delta_W - \frac{1}{2} + \Lambda_2(k^2, M_W)\right] \\ &\quad - ieI_\sigma^3\gamma_\mu(1-\gamma_5)\frac{\alpha}{4\pi}\frac{3}{2s^2}\left[\Delta_W - \frac{1}{6} + \Lambda_3(k^2, M_W)\right]^{8)},\end{aligned} \tag{5.28}$$

[7]) In the following we drop the fermion family index i since Cabbibo rotation is not involved.
[8]) In eqs. (5.28, 30) σ' denotes the isospin partner of the fermion σ.

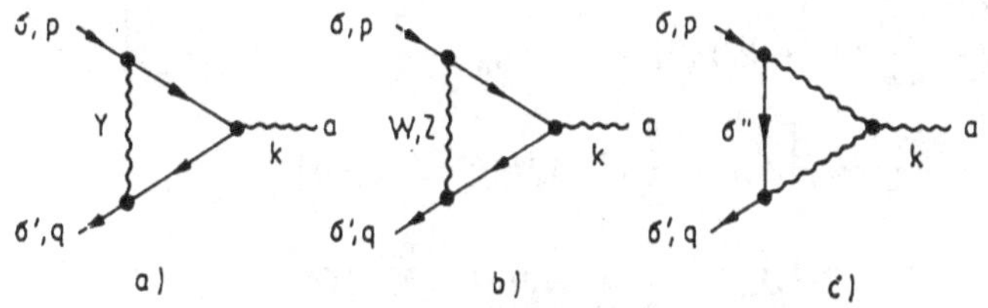

Fig. 7. 1-loop gauge boson fermion vertex diagrams

$$\begin{aligned} \Gamma_\mu^{\gamma\nu\nu}(k^2) &= ie\gamma_\mu(1-\gamma_5)\,\frac{\alpha}{4\pi}\,\frac{1}{4s^2}\left[\Delta_W - \frac{1}{2} + \Lambda_2(k^2, M_W)\right] \\ &\quad - ie\gamma_\mu(1-\gamma_5)\,\frac{\alpha}{4\pi}\,\frac{3}{4s^2}\left[\Delta_W - \frac{1}{6} + \Lambda_3(k^2, M_W)\right], \end{aligned} \tag{5.29}$$

$$\begin{aligned} \Gamma_\mu^{Z\sigma\sigma}(k^2) &= ie\gamma_\mu(v_\sigma - a_\sigma\gamma_5) + ie\gamma_\mu(v_\sigma - a_\sigma\gamma_5) \\ &\quad \times \frac{\alpha}{4\pi}\,Q_\sigma^2\left[\Delta_0 - 2\ln\frac{m_\sigma^2}{\lambda^2} + 4 + \Lambda_1(k^2, m_\sigma)\right] + ie[v_\sigma(v_\sigma^2 + 3a_\sigma^2)\,\gamma_\mu \\ &\quad - a_\sigma(3v_\sigma^2 + a_\sigma^2)\,\gamma_\mu\gamma_5]\,\frac{\alpha}{4\pi}\left[\Delta_Z - \frac{1}{2} + \Lambda_2(k^2, M_Z)\right] \\ &\quad + ie\gamma_\mu(1-\gamma_5)\,\frac{\alpha}{4\pi}\,\frac{v_{\sigma'} + a_{\sigma'}}{4s^2}\left[\Delta_W - \frac{1}{2} + \Lambda_2(k^2, M_W)\right] \\ &\quad + ieI_\sigma^3\gamma_\mu(1-\gamma_5)\,\frac{\alpha}{4\pi}\,\frac{3c}{2s^3}\left[\Delta_W - \frac{1}{6} + \Lambda_3(k^2, M_W)\right], \end{aligned} \tag{5.30}$$

$$\begin{aligned} \Gamma_\mu^{Z\nu\nu}(k^2) &\doteq i\,\frac{e}{2sc}\,\gamma_\mu\,\frac{1-\gamma_5}{2}\left\{1 + \frac{\alpha}{4\pi}\,\frac{1}{4s^2c^2}\left[\Delta_Z - \frac{1}{2} + \Lambda_2(k^2, M_Z)\right]\right. \\ &\quad + \frac{\alpha}{4\pi}\,\frac{2s^2-1}{2s^2}\left[\Delta_W - \frac{1}{2} + \Lambda_2(k^2, M_W)\right] \\ &\quad \left. + \frac{\alpha}{4\pi}\,\frac{3c^2}{5^2}\left[\Delta_W - \frac{1}{6} + \Lambda_3(k^2, M_W)\right]\right\}, \end{aligned} \tag{5.31}$$

$$\begin{aligned} \Gamma_\mu^{We\nu}(k^2) &= i\,\frac{e}{\sqrt{2}\,s}\,\gamma_\mu\,\frac{1-\gamma_5}{2}\left\{1 + \frac{\alpha}{4\pi}\cdot\frac{2s^2-1}{4s^2c^2}\left[\Delta_Z - \frac{1}{2} + \Lambda_2(k^2, M_Z)\right]\right. \\ &\quad + \frac{\alpha}{4\pi}\cdot 3\left[\Delta_W + \frac{5}{6} + \Lambda_4^{m_e}(k^2; M_W, 0)\right] \\ &\quad \left. + \frac{\alpha}{4\pi}\cdot\frac{3c^2}{s^2}\left[\Delta_W + \frac{5}{6} + \frac{M_Z^2}{M_Z^2 - M_W^2}\ln\frac{M_W^2}{M_Z^2} + \Lambda_4(k^2; M_Z, M_W)\right]\right\}, \end{aligned} \tag{5.32}$$

$$\begin{aligned} \Gamma_\mu^{Wdu}(k^2) &= \frac{ie}{2\sqrt{2}\,s}\,\gamma_\mu(1-\gamma_5)\cdot\left\{1 + \frac{\alpha}{4\pi}\,Q_uQ_d\left[\Delta_d - 2\ln\frac{m_d^2}{\lambda^2} + 4 + 3\ln\frac{m_d}{m_u}\right.\right. \\ &\quad \left. + \frac{1}{2}\,\Lambda_1(k^2, m_d) + \frac{1}{2}\,\Lambda_1(k^2, m_u)\right] \end{aligned}$$

$$+ \frac{\alpha}{4\pi} (v_u + a_u)(v_d + a_d) \left[\Delta_Z - \frac{1}{2} + \Lambda_2(k^2, M_Z) \right]$$

$$+ \frac{\alpha}{4\pi} 3Q_u \left[\Delta_W + \frac{5}{6} + \Lambda_4^{m_u}(k^2; M_W, 0) \right]$$

$$- \frac{\alpha}{4\pi} 3Q_d \left[\Delta_W + \frac{5}{6} + \Lambda_4^{m_d}(k^2, M_W, 0) \right]$$

$$+ \frac{\alpha}{4\pi} \frac{3c^2}{s^2} \left[\Delta_W + \frac{5}{6} + \frac{M_Z^2}{M_Z^2 - M_W^2} \ln \frac{M_W^2}{M_Z^2} + \Lambda_4(k^2; M_Z, M_W) \right] \Bigg\}.$$

The invariant functions $\Lambda_{1,\dots,4}$ together with some of their properties are presented in app. B.3.

For the renormalization of the electric charge we need the γee-vertex at $k^2 = 0$, $p^2 = q^2 = m_e^2$. Its explicit form is:

$$\Gamma_\mu^{\gamma ee}(0) = ie\gamma_\mu + ie\gamma_\mu \frac{\alpha}{4\pi} \left[\Delta_e - 2 \ln \frac{m_e^2}{\lambda^2} + 4 \right]$$

$$+ ie \frac{\alpha}{4\pi} [(v_e^2 + a_e^2)\gamma_\mu - 2v_e a_e \gamma_\mu \gamma_5] \left[\Delta_Z - \frac{1}{2} \right]$$

$$+ ie\gamma_\mu \frac{1 - \gamma_5}{2} \frac{\alpha}{4\pi} \frac{3}{2s^2} \left[\Delta_W - \frac{1}{6} \right]. \tag{5.33}$$

For later use we give also the neutrino-photon vertex at $k^2 = 0$:

$$\Gamma_\mu^{\gamma\nu\nu} = -ie\gamma_\mu \frac{1 - \gamma_5}{2} \frac{\alpha}{4\pi} \frac{1}{s^2} \Delta_W. \tag{5.34}$$

5.3 Renormalization constants

The prescriptions for the calculation of the renormalization constants from the unrenormalized self energies and vertex functions have been defined in eq.s (4.1′) to (4.6′). We find for the mass renormalization $\delta \overline{M}_W^2$, $\delta \overline{M}_Z^2$ of the heavy gauge bosons

$$\delta \overline{M}_W^2 = \delta \overline{M}_{W,\text{sing}}^2 + \delta \overline{M}_{W,\text{fin}}^2 = \Sigma_{T,\text{sing}}^W(M_W^2) + \text{Re}\, \Sigma_{T,\text{fin}}^W(M_W^2), \tag{5.35}$$

$$\delta \overline{M}_Z^2 = \delta \overline{M}_{Z,\text{sing}}^2 + \delta \overline{M}_{Z,\text{fin}}^2 = \Sigma_{T,\text{sing}}^Z(M_Z^2) + \text{Re}\, \Sigma_{T,\text{fin}}^Z(M_Z^2) \tag{5.36}$$

explicit expressions by using eq.s (5.9) and (5.10). Eq.s (4.1′) tell us that with $\delta \overline{M}_W^2$, $\delta \overline{M}_Z^2$ also the following combination of δe, δZ_1^Z, δZ_2^Z is determined:

$$\frac{\delta \overline{M}_Z^2}{M_Z^2} - \frac{\delta \overline{M}_W^2}{M_W^2} = \frac{s^2}{c^2 - s^2} \left(2 \frac{\delta e}{e} - 2\delta Z_1^Z + 3\delta Z_2^Z \right). \tag{5.37}$$

Eq.s (4.2′) together with (5.7), (5.8), (5.33) give for the photon field renormalization constant

$$\delta Z_2^\gamma = \frac{\alpha}{4\pi} \left[-\frac{4}{3} \sum_{i\sigma} (Q_{i\sigma}^2 \Delta_{i\sigma}) + 3\Delta_W + \frac{2}{3} \right], \tag{5.38}$$

a combination of the (γ, Z) renormalization constants

$$\delta Z_1^{\gamma Z} - \delta Z_2^{\gamma Z} = \frac{sc}{c^2 - s^2} (\delta Z_1^Z - \delta Z_2^Z - \delta Z_1^\gamma + \delta Z_2^\gamma) = -\frac{\alpha}{4\pi} \frac{2c}{s} \Delta_W \quad (5.39)$$

and the charge renormalization

$$\frac{\delta e}{e} = \delta Z_1^\gamma - \frac{3}{2} \delta Z_2^\gamma = \frac{\alpha}{4\pi} \left[\frac{2}{3} \Sigma_{i\sigma}(Q_{i\sigma}^2 \Delta_{i\sigma}) - \frac{7}{2} \Delta_W - \frac{1}{3}\right]. \quad (5.40)$$

A comparison of eq. (5.40) with (5.38) shows that

$$\frac{\delta e}{e} = -\frac{1}{2} \delta Z_2^\gamma - \frac{\alpha}{4\pi} \cdot 2\Delta_W. \quad (5.41)$$

This means that the familiar QED relation is modified by the non-Abelian couplings of the gauge bosons.

The four eq.s (5.37) to (5.40) allow the separate determination of $\delta Z_1^\gamma, \delta Z_2^\gamma, \delta Z_1^Z, \delta Z_2^Z$:

$$\delta Z_1^\gamma = \frac{\alpha}{4\pi} \left[-\frac{4}{3} \sum_{i\sigma} (Q_{i\sigma}^2 \Delta_{i\sigma}) + \Delta_W + \frac{2}{3}\right],$$

$$\delta Z_2^\gamma = \frac{\alpha}{4\pi} \left[-\frac{4}{3} \sum_{i\sigma} (Q_{i\sigma}^2 \Delta_{i\sigma}) + 3\Delta_W + \frac{2}{3}\right],$$

$$\delta Z_1^Z = \frac{\alpha}{4\pi} \left[-\frac{4}{3} \sum_{i\sigma} (Q_{i\sigma}^2 \Delta_{i\sigma}) + \left(7 - 6\frac{c^2}{s^2}\right) \Delta_W + \frac{2}{3}\right] + \frac{c^2 - s^2}{s^2} \left(\frac{\delta M_Z^2}{M_Z^2} - \frac{\delta M_W^2}{M_W^2}\right), \quad (5.42)$$

$$\delta Z_2^Z = \frac{\alpha}{4\pi} \left[-\frac{4}{3} \sum_{i\sigma} (Q_{i\sigma}^2 \Delta_{i\sigma}) + \left(7 - 4\frac{c^2}{s^2}\right) \Delta_W + \frac{2}{3}\right] + \frac{c^2 - s^2}{s^2} \left(\frac{\delta M_Z^2}{M_Z^2} - \frac{\delta M_W^2}{M_W^2}\right).$$

Together with these constants also $\delta Z_1^W, \delta Z_2^W, \delta Z_1^{\gamma Z}, \delta Z_2^{\gamma Z}$ are determined. Explicit expressions may be obtained with help of eq. (A.1).

The mass and field renormalization of the leptons according to eq.s (4.1′) and (4.3′) treats the charged, massive leptons and the neutral, massless neutrinos in an unsymmetric way. This is a consequence of spontaneous breaking of $SU(2) \times U(1)$ and of chiral symmetry. As a result the neutrinos remain massless and left-handed after renormalization, whereas the charged leptons suffer mass renormalization:

$$\frac{\delta m_{i\sigma}}{m_{i\sigma}} = \frac{\alpha}{4\pi} \left\{\frac{1}{4s^2} \left(\Delta_W - \frac{1}{2}\right) + 2 (v_{i\sigma}^2 + a_{i\sigma}^2) \left(\Delta_Z - \frac{1}{2}\right) - 4(v_{i\sigma}^2 - a_{i\sigma}^2) \left(\Delta_Z + \frac{1}{2}\right) - 3\left(\Delta_{i\sigma} + \frac{4}{3}\right)\right\} = -\frac{\delta v}{v} + \delta Z_1^{i\sigma}. \quad (5.43)$$

The residue of the electron propagator was put equal to one for both the L and R parts. This gives:

$$\delta Z_L^{(e,\nu)} = -\frac{\alpha}{4\pi}\left\{\Delta_e - 2\ln\frac{m_e^2}{\lambda^2} + 4 + (v_e + a_e)^2\left(\Delta_Z - \frac{1}{2}\right) + \frac{1}{2s^2}\left(\Delta_W - \frac{1}{2}\right)\right\},$$

$$\delta Z_R^e = -\frac{\alpha}{4\pi}\left\{\Delta_e - 2\ln\frac{m_e^2}{\lambda^2} + 4 + (v_e - a_e)^2\left(\Delta_Z - \frac{1}{2}\right)\right\}. \tag{5.44}$$

SAKAKIBARA [12] applies the condition Res $= 1$ to the ν propagator which consequently does not get an artificial IR singularity. But in this scheme the electronfield is not treated like in QED.

In the case of quarks we have two right-handed singlets associated with one left-handed doublet. The two mass renormalization constants are determined by eq. (5.43). The doublet renormalization constant $\delta Z_L^{(d,u)}$ and the singlet renormalization constant δZ_R^d for the $I_3 = -1/2$ members are fixed as in the lepton case: L and R residues in the d-propagator are put equal to one.

δZ_R^u is determined such that the residues of the L and R parts in the u-propagator are still equal (but $\neq 1$). This yields:

$$\delta Z_L^{(d,u)} = -\frac{\alpha}{4\pi}\left\{Q_d^2\left(\Delta_d - 2\ln\frac{m_d^2}{\lambda^2} + 4\right) + (v_d + a_d)^2\left(\Delta_Z - \frac{1}{2}\right) + \frac{1}{2s^2}\left(\Delta_W - \frac{1}{2}\right)\right\},$$

$$\delta Z_R^d = -\frac{\alpha}{4\pi}\left\{Q_d^2\left(\Delta_d - 2\ln\frac{m_d^2}{\lambda^2} + 4\right) + (v_d - a_d)^2\left(\Delta_Z - \frac{1}{2}\right)\right\}, \tag{5.45}$$

$$\delta Z_R^u = -\frac{\alpha}{4\pi}\left\{Q_u^2\left(\Delta_u - 2\ln\frac{m_u^2}{\lambda^2} + 4\right) + (v_u - a_u)^2\left(\Delta_Z - \frac{1}{2}\right) - \delta(u, d)\right\},$$

with

$$\delta(u, d) = Q_u^2\left(\ln\frac{M_Z^2}{m_u^2} - 2\ln\frac{m_u^2}{\lambda^2}\right) - Q_d^2\left(\ln\frac{M_Z^2}{m_d^2} - 2\ln\frac{m_d^2}{\lambda^2}\right) + \frac{3}{2}. \tag{5.46}$$

Since the propagators of the u-type fermions have a residue different from 1 a wave function renormalization factor $1 - \alpha/4\pi \cdot \delta(u, d)/2$ has to be assigned to each external line.

The Higgs mass M_H and the Higgs field are renormalized using the prescriptions (4.1′) and (4.3′) for δM_H^2 and δZ^φ together with the expression (5.18) for the unrenormalized Higgs self energy. This gives:

$$\left.\frac{\delta M_H^2}{M_H^2}\right|_{\text{sing}} = \frac{\alpha}{4\pi}\Delta_W\left(\frac{3 + 6c^2}{8c^2s^2} - \frac{17}{4c^2s^2}\frac{M_Z^2 + 2c^2M_W^2}{M_H^2} - \frac{27}{8s^2}\frac{M_H^2}{M_W^2}\right),$$

$$\left.\frac{\delta M_H^2}{M_H^2}\right|_{\text{fin}} = \frac{\alpha}{4\pi}\left\{-\frac{2c^2 + 1}{8c^2s^2} + \frac{7}{4c^2s^2}\frac{M_Z^2 + 2c^2M_W^2}{M_H^2} - \frac{3}{8s^2}\frac{M_H^2}{M_W^2}\right. \tag{5.47}$$

$$\left. + \frac{1}{4c^2s^2}\left(-\frac{3}{2} + 17\frac{M_Z^2}{M_H^2} + \frac{M_H^2}{M_Z^2}\right)\ln\frac{M_Z^2}{M_W^2} + \frac{21}{8s^2}\frac{M_H^2}{M_W^2}\ln\frac{M_H^2}{M_W^2}\right.$$

$$+ \frac{1}{s^2}\left(1 - 7\frac{M_W^2}{M_H^2} - \frac{M_H^2}{2M_W^2}\right) \operatorname{Re} F(M_H^2; M_W, M_W)$$

$$+ \frac{1}{2c^2s^2}\left(1 - 7\frac{M_Z^2}{M_H^2} - \frac{M_H^2}{2M_W^2}\right) \operatorname{Re} F(M_H^2; M_Z, M_Z) - \frac{9}{4s^2}\frac{M_H^2}{M_W^2}\Bigg\};$$

$$\delta Z^\varphi_{\text{sing}} = \frac{\alpha}{4\pi} \Delta_W \frac{2c^2+1}{2c^2s^2},$$

$$\delta Z^\varphi_{\text{fin}} = -\frac{\alpha}{4\pi} \operatorname{Re}\left\{\frac{1}{2c^2s^2}\ln\frac{M_Z^2}{M_W^2} - \frac{1}{s^2} F(M_H^2; M_W, M_W)\right. \tag{5.48}$$

$$- \frac{1}{2c^2s^2} F(M_H^2; M_Z, M_Z) + \frac{1}{s^2}\left(7M_W^2 - M_H^2 + \frac{M_H^4}{2M_W^2}\right)$$

$$\times F'(M_H^2; M_W, M_W) + \frac{1}{2c^2s^2}\left(7M_Z^2 - M_H^2 + \frac{M_H^4}{2M_Z^2}\right) F'(M_H^2; M_Z, M_Z)$$

$$\left. + \frac{9}{4s^2}\frac{M_H^4}{M_W^2} F'(M_H^2; M_H, M_H)\right\}.$$

The renormalization of the gauge fixing parameters ξ follows from eq. (4.5′) together with the expressions (5.11—14), (5.19, 20) for the longitudinal parts of the gauge boson self energies and the self energies of the unphysical Higgs fields:

$$\delta\xi_1^\gamma = \delta Z_2^\gamma, \tag{5.49}$$

$$\delta\xi_1^Z = -\frac{\alpha}{4\pi}\frac{1}{c^2s^2} \operatorname{Re}\left\{(2c^2 - 4c^2s^2 - 1)\Delta_W - 1\right.$$

$$+ \frac{1}{M_H^2 - M_Z^2}\left(M_H^2 \ln\frac{M_H^2}{M_W^2} - M_Z^2 \ln\frac{M_Z^2}{M_W^2}\right)$$

$$+ 2c^2(c^2 - s^2) F(M_Z^2; M_W, M_W)$$

$$\left. - \left(1 - \frac{(M_H^2 - M_Z^2)^2}{4M_Z^4}\right) F(M_Z^2; M_H, M_Z)\right\} - \frac{\delta M_Z^2}{M_Z^2}, \tag{5.50}$$

$$\delta\xi_1^W = -\frac{\alpha}{4\pi}\frac{1}{c^2s^2} \operatorname{Re}\left\{(c^2 - s^2)(\Delta_W + 1)\right.$$

$$+ \left(3 - \frac{2}{s^2}\right)\ln\frac{M_Z^2}{M_W^2} + c^2\frac{M_H^2}{(M_H^2 - M_W^2)}\ln\frac{M_H^2}{M_W^2}$$

$$+ 2c^2s^2 F(M_W^2; 0, M_W) + \left(2s^4c^2 + 3c^2 - 1 + \frac{s^4}{4}\right) F(M_W^2; M_W, M_Z)$$

$$\left. + c^2\left(\frac{(M_H^2 - M_W^2)^2}{4M_W^4} - 1\right) F(M_W^2; M_W, M_H)\right\} - \frac{\delta M_W^2}{M_W^2}, \tag{5.51}$$

$$\delta\xi_2^{\gamma Z} = \frac{cs}{c^2 - s^2}(\delta\xi_1^Z - \delta\xi_1^\gamma) + \frac{\alpha}{4\pi}\cdot 4\frac{c}{s}\Delta_W,$$

$$\delta\xi_2^Z = -\frac{\alpha}{4\pi}\frac{1}{2c^2s^2}\,\mathrm{Re}\left\{(2c^2+1)\,\Delta_W + 1 - \frac{M_H^2}{M_H^2 - M_Z^2}\ln\frac{M_H^2}{M_W^2}\right.$$
$$+ \frac{M_Z^2}{M_H^2 - M_Z^2}\ln\frac{M_Z^2}{M_W^2} + 2c^2F(M_Z^2; M_W, M_W)$$
$$\left. + \left(1 - \frac{(M_H^2 - M_Z^2)^2}{2M_Z^4}\right)F(M_Z^2; M_H, M_Z)\right\} - \frac{\delta M_Z^2}{M_Z^2}, \tag{5.52}$$

$$\delta\xi_2^W = -\frac{\alpha}{4\pi}\frac{1}{2c^2s^2}\,\mathrm{Re}\left\{(2c^2+1)\,\Delta_W + 2c^2 + 1 + \left(4c^2 - \frac{1+c^2}{s^2}\right)\ln\frac{M_Z^2}{M_W^2}\right.$$
$$+ c^2\frac{M_H^2}{M_W^2 - M_H^2}\ln\frac{M_H^2}{M_W^2} + \frac{1}{2}\left(5 - 9s^2 - \frac{1}{s^2}\right)F(M_W^2; M_W, M_Z)$$
$$\left. + c^2\left(1 - \frac{(M_W^2 - M_H^2)^2}{2M_W^4}\right)F(M_W^2; M_W, M_H)\right\} - \frac{\delta M_W^2}{M_W^2}. \tag{5.53}$$

Finally we have to renormalize the ghost fields. Eq.s (4.6′) and (5.21, 22) give:

$$\delta\tilde{Z}^\gamma = s^2\delta\tilde{Z}^W + c^2\delta\tilde{Z}^B = \frac{1}{2}\,\delta Z_2^\gamma + \frac{\alpha}{4\pi}\,\Delta_W, \tag{5.54}$$

$$\delta\tilde{Z}^{\gamma Z} = cs(\delta\tilde{Z}^W - \delta\tilde{Z}^B) = -\delta Z_1^{\gamma Z} + \frac{3}{2}\,\delta Z_2^{\gamma Z} - \frac{\alpha}{4\pi}\frac{c}{s}\,\Delta_W. \tag{5.55}$$

With these expressions we have determined all the renormalization constants of our renormalization scheme of the standard model. They can be used together with the counterterms to derive the finite renormalized Green functions of the model in 1-loop order. In the next section we present the results for the self energies of the gauge bosons and the fermions as well as the fermion gauge boson vertices and box diagrams.

6. Renormalized Self Energies and Vertex Functions

6.1 Gauge boson self energies

In order to give an impression of the influence of the 1-loop contributions on the magnitude of radiative corrections we present in this section the formulas for the renormalized self energies and vertex functions and numerical results for these quantities. From the expressions (5.7—14) for the unrenormalized self energies and the renormalization constants (5.35—42) together with the prescriptions for the renormalization we obtain for the renormalized transverse parts of the gauge boson self energies the following formulas:

$$\hat{\Sigma}_T^\gamma(k^2) = \Sigma_T^\gamma(k^2) - k^2\frac{\partial}{\partial k^2}\Sigma_T^\gamma(k^2)|_{k^2=0},$$

$$\hat{\Sigma}_T^{\gamma Z}(k^2) = \Sigma_T^{\gamma Z}(k^2)|_{\mathrm{fin}}$$
$$- k^2\frac{c}{s}\,\mathrm{Re}\left(\frac{\Sigma_T^Z(M_Z^2)}{M_Z^2} - \frac{\Sigma_T^W(M_W^2)}{M_W^2} + \frac{\alpha}{4\pi}\frac{m_t^2 - m_b^2}{4s^2M_W^2}\ln\frac{m_t^2}{m_b^2}\right)\Bigg|_{\mathrm{fin}}$$
$$- k^2\frac{1}{6cs}\,\delta_q, \tag{6.1}$$

$$\hat{\Sigma}_T{}^Z(k^2) = \left(\Sigma_T{}^Z(k^2) - \mathrm{Re}\,\Sigma_T{}^Z(M_Z{}^2)\right)\Big|_{\text{fin}}$$
$$+ (k^2 - M_Z{}^2)\left(\frac{\alpha}{6\pi} + \frac{c^2 - s^2}{s^2}\,\mathrm{Re}\left(\frac{\Sigma_T{}^Z(M_Z{}^2)}{M_Z{}^2} - \frac{\Sigma_T{}^W(M_W{}^2)}{M_W{}^2}\right.\right.$$
$$\left.\left. + \frac{\alpha}{4\pi}\,\frac{m_t{}^2 - m_b{}^2}{4s^2M_W{}^2}\ln\frac{m_t{}^2}{m_b{}^2}\right)\Big|_{\text{fin}} + \frac{1}{3s^2}\,\delta_q\right),$$

$$\hat{\Sigma}_T{}^W(k^2) = \left(\Sigma_T{}^W(k^2) - \mathrm{Re}\,\Sigma_T{}^W(M_W{}^2)\right)\Big|_{\text{fin}}$$
$$+ (k^2 - M_W{}^2)\left(\frac{\alpha}{6\pi} + \frac{c^2}{s^2}\,\mathrm{Re}\left(\frac{\Sigma_T{}^Z(M_Z{}^2)}{M_Z{}^2} - \frac{\Sigma_T{}^W(M_W{}^2)}{M_W{}^2}\right.\right.$$
$$\left.\left. + \frac{\alpha}{4\pi}\,\frac{m_t{}^2 - m_b{}^2}{4s^2M_W{}^2}\ln\frac{m_t{}^2}{m_b{}^2}\right)\Big|_{\text{fin}} + \frac{1}{3s^2}\,\delta_q\right),$$

with

$$\delta_q = \frac{\alpha}{4\pi}\left(\ln\frac{m_u{}^2}{m_d{}^2} + \ln\frac{m_c{}^2}{m_s{}^2} + \ln\frac{m_t{}^2}{m_b{}^2}\right).$$

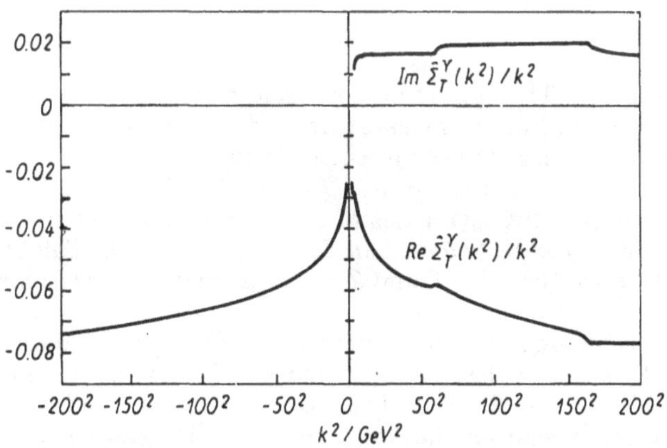

Fig. 8. Real and imaginary parts of the renormalized transverse photon self energy $\hat{\Sigma}_T{}^\gamma(k^2)$. The curve shows $\hat{\Sigma}_T{}^\gamma(k^2)/k^2$

The numerical results in figs. 8—11 for $|k^2| < (200\ \text{GeV})^2$ have been calculated with the following standard set of parameters (if possible taken from [26]):

$$\begin{array}{llll}
\alpha = (137.036)^{-1}, & & & \\
M_W = 82\ \text{GeV}, & M_Z = 93\ \text{GeV}, & M_H = 100\ \text{GeV}, & \\
m_e = 0.511\ \text{MeV}, & m_\mu = 105.66\ \text{MeV}, & m_\tau = 1784\ \text{MeV}, & (6.2) \\
m_u = 5\ \text{MeV}, & m_d = 7\ \text{MeV}, & m_s = 150\ \text{MeV}\ [27], & \\
m_c = 1.5\ \text{GeV}, & m_b = 4.5\ \text{GeV}, & m_t = 30\ \text{GeV}. &
\end{array}$$

The real parts of the diagonal self energies $\hat{\Sigma}_T{}^\gamma$, $\hat{\Sigma}_T{}^Z$, $\hat{\Sigma}_T{}^W$ are, compared to the free inverse propagators, not small but yield 10% as typical order of magnitude. The contributions to the imaginary parts from the fermion loops are positive but those from the gauge loops are negative. Both the real and imaginary parts of the W and Z self energy

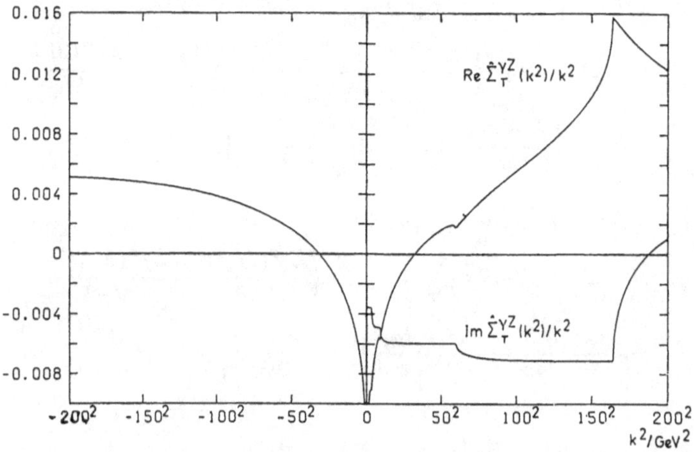

Fig. 9. Photon Z boson mixing. Presented is $\hat{\Sigma}_T{}^{\gamma Z}(k^2)/k^2$

depend strongly on the energy. The approximation using a constant imaginary part in the vicinity of the resonance leading to the usual Breit-Wigner type form of the modulus square of the propagator is thus not really justified. In fig. 12, 13 we show a comparison between a Breit-Wigner distribution using M_W and Im $\hat{\Sigma}_T{}^W(M_W{}^2)$ resp. M_Z and Im $\hat{\Sigma}_T{}^Z(M_Z{}^2)$ and the corresponding quantities resulting from (6.1). We find for the W and Z FWHM values which are 10% bigger than Im $\hat{\Sigma}_T(M^2)/M = \Gamma$. This means that for the determination of the width of the W and Z a careful analysis of the experimental distributions is necessary.

In the case of the W self energy we have contributions of loops containing photons. The physical channel $W \to W + \gamma$ has its threshold at $k^2 = M_W{}^2$. Consequently we observe in fig. 11 the peak in the real part and the structure in the imaginary part. The magnitudes of these effects depend on the details of the $WW\gamma$ coupling. In a model

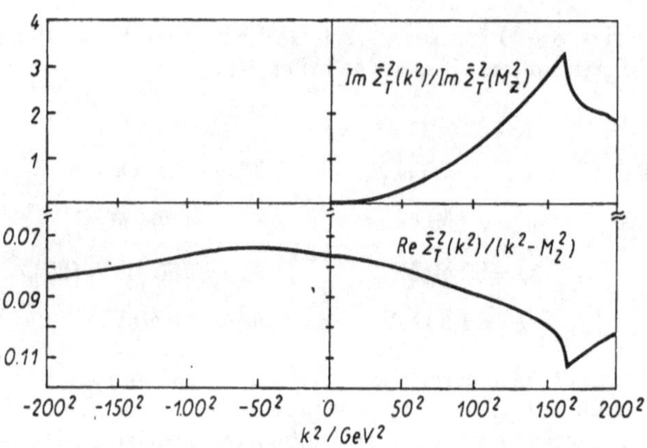

Fig. 10

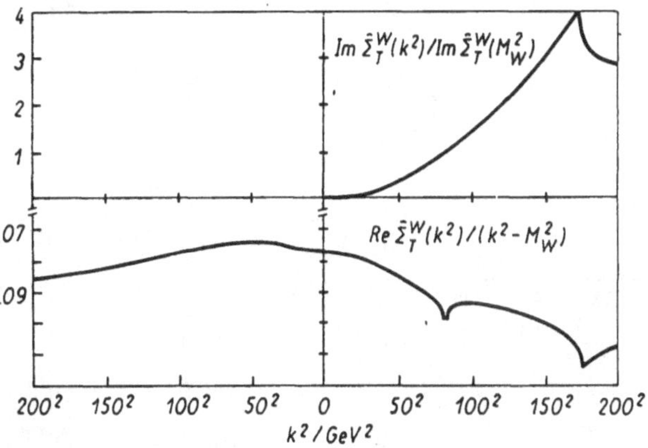

Fig. 11

Fig. 10, 11. Renormalized W, Z self energy. The curves are $\mathrm{Re}\,\hat{\Sigma}_T^{W,Z}(k^2)/(k^2 - M_{W,Z}^2)$ and $\mathrm{Im}\,\hat{\Sigma}_T^{W,Z}(k^2)/\mathrm{Im}\,\Sigma_T^{W,Z}(M_{W,Z}^2)$ (Read −0.06, −0.08, −0.10 instead of −0.07, −0.09, −0.11, resp.)

where the W is coupled to the photon in the form of a minimal substitution it would be different from that of the standard model.

The diagonal gauge boson self energies are very large compared to α/π and therefore will give the main contributions besides bremsstrahlung to the radiative corrections in e^+e^- annihilation. Compared to these the γZ mixing is much smaller and in our renormalization scheme typically of the order of magnitude of 1%. In our scheme we do not use the weak mixing angle θ_W as a fundamental parameter but as a short hand for $\sin^2\theta_W = (1 - M_W{}^2/M_Z{}^2)$. The results shown in fig. 9 might be interpreted as contributions to an effective running i.e. energy depending mixing angle.

The residue of the renormalized Z propagator is different from 1. We obtain:

$$\frac{\partial}{\partial k^2}\hat{\Sigma}_T{}^Z(k^2)|_{k^2=M_Z{}^2} = -0.080 + i\,0.029. \tag{6.3}$$

For comparison with other renormalization schemes we present also $\hat{\Sigma}_T{}^W(0)$ and $\hat{\Sigma}_T{}^Z(0)$:

$$\begin{aligned}&\frac{\partial}{\partial k^2}\hat{\Sigma}_T{}^{\gamma Z}(k^2)|_{k^2=0} = \Pi^{\gamma Z}(0) = -0.021\\ &\hat{\Sigma}_T{}^W(0)/M_W{}^2 = -\Pi^W(0) = 0.069, \qquad \hat{\Sigma}_T{}^Z(0)/M_Z{}^2 = -\Pi^Z(0) = 0.069.\end{aligned} \tag{6.4}$$

These values enter into the calculation of radiative corrections to low energy processes.

The parameters (6.2) which we used for the numerical discussion of the self energies are only partly known from experiment. The W and Z masses have been chosen in agreement with the $P\bar{P}$-collider results [1] but since they still have rather large errors we have investigated the dependence of the self energies on the masses M_W, M_Z of the gauge bosons. We find that it is determined mainly through the ratio M_W/M_Z. Therefore we present as an example in Fig.s 14 and 15 $\Pi^W(0)$, $\Pi^{\gamma Z}(0)$ as functions of $s_W{}^2 = 1 - M_W{}^2/M_Z{}^2$ and M_W.

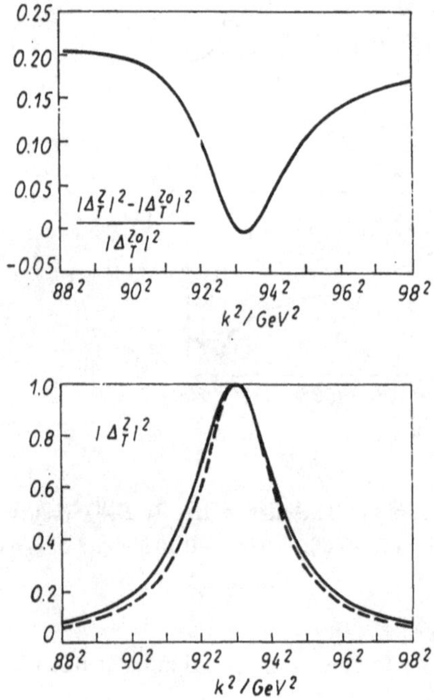

Fig. 12

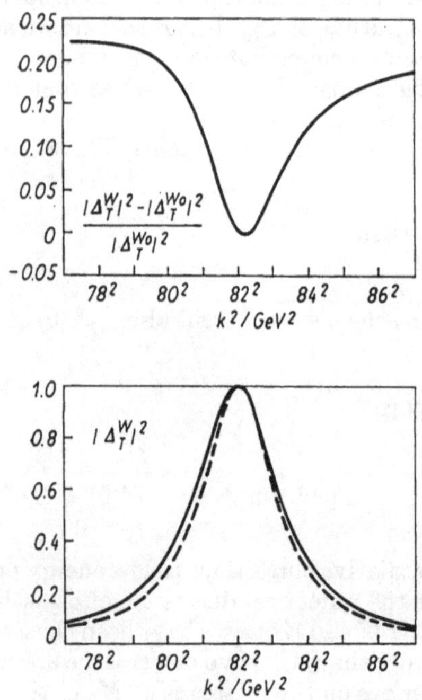

Fig. 13

Fig. 12, 13. Comparison between a Breit-Wigner distribution with $M_{W,Z}$ and Im $\hat{\Sigma}_T^{W,Z}$ $\times$ $(M^2_{W,Z})$ as parameters (- - - - -) and the square of the modulus of the renormalized W, Z propagators (———)

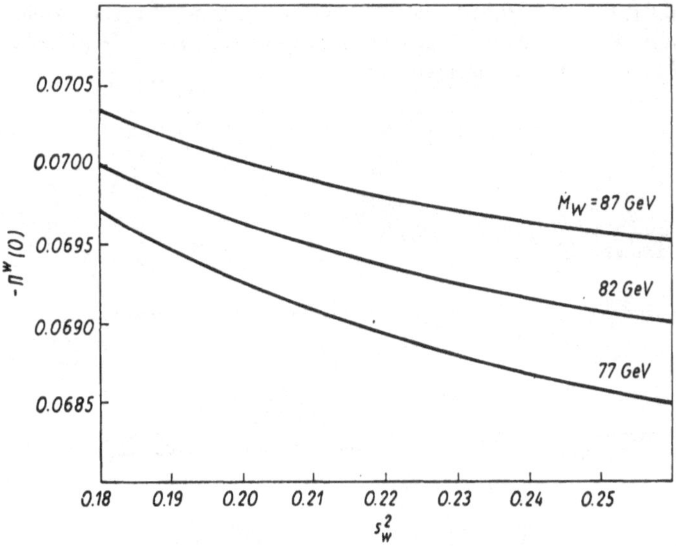

Fig. 14. The W self energy $-\Pi^W(0) = \hat{\Sigma}_T{}^W(0)/M_W{}^2$ as function of $s_W{}^2$, M_W for the standard set of parameters (6.2)

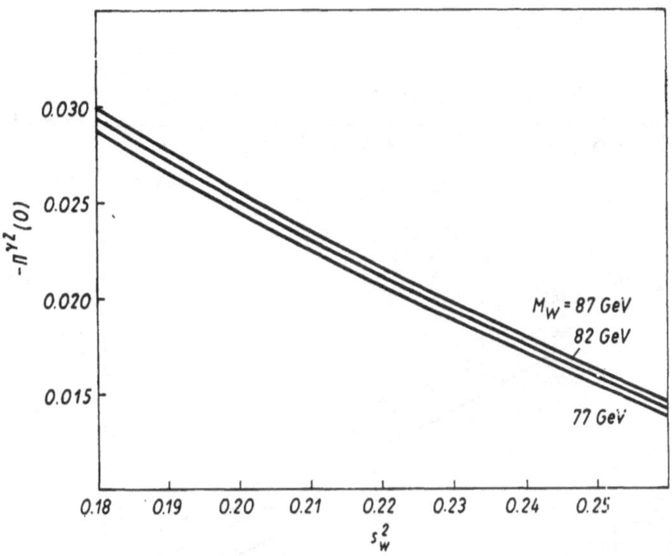

Fig. 15. The γZ mixing $\Pi^{\gamma Z}(0)$ as function of $s_W{}^2$, M_W (other parameters like in fig. 14)

The masses for the u, d, s quarks used above correspond to the values obtained by GASSER and LEUTWYLER [27]. The evaluation of the hadronic contribution to the photon self energy using quark loops with these values of the quark masses leads for energies between 10 and 100 GeV to numerical results which are in agreement with a determination [28] of these contribution using dispersion theory and experimental data from e^+e^--annihilation. However, since the quark masses are not known very precisely and since in the literature calculations of radiative corrections using much bigger values for these masses can be found, we have studied the dependence of Π^W, $\Pi^{\gamma Z}$, Π^Z on $m_i = \{m_u, m_d, m_s\}$. Defining $\delta_q \Pi = \Pi(m_{i_1}) - \Pi(m_{i_2})$ we find:

$$\begin{aligned} \delta_q \Pi^W &= \delta_q \Pi^Z = -\frac{2\alpha}{\pi} \sum_i Q_i^2 \ln \frac{m_{i_1}}{m_{i_2}}, \\ \delta_q \Pi^{\gamma Z} &= \frac{\alpha}{4\pi c_W s_W} \sum_i Q_i (I_i^3 - 2Q_i) \ln \frac{m_{i_1}}{m_{i_2}}. \end{aligned} \tag{6.5}$$

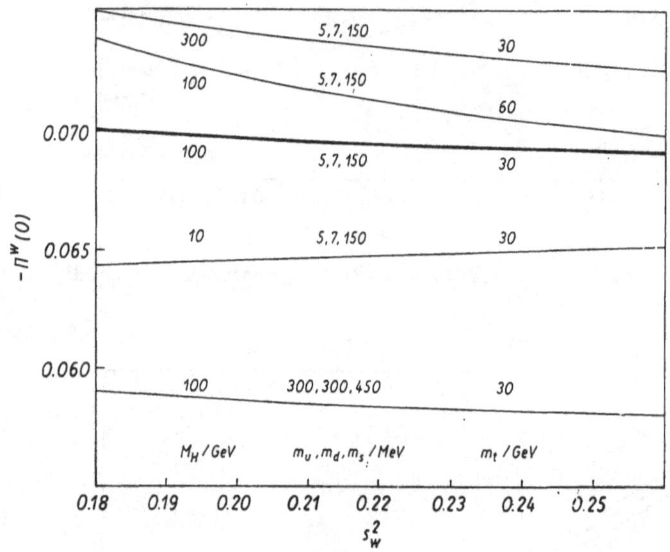

Fig. 16 a

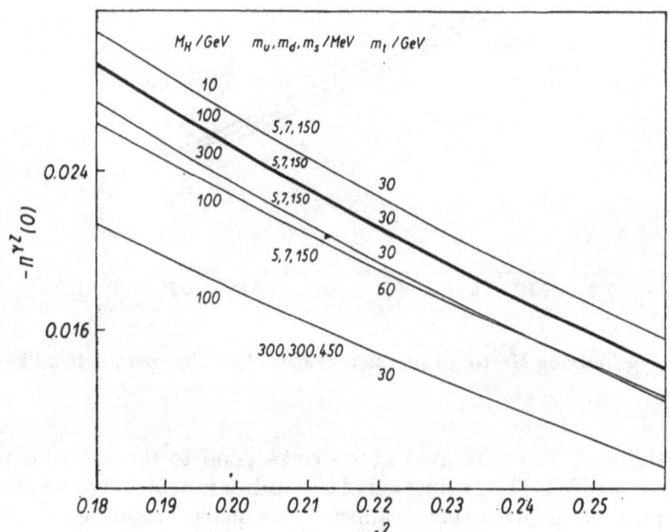

Fig. 16 b

Fig. 16a, b. $\Pi^W(0)$ (a), $\Pi^{\gamma Z}(0)$ (b), as functions of s_W^2 with $M_W = 82$ GeV fixed. Shown are the variations with the Higgs mass, top quark mass and the masses of the light quarks

The extreme choice $m_u = m_d = 300$ MeV, $m_s = 450$ MeV leads to the curve for $\Pi^W(0)$, $\Pi^{\gamma Z}(0)$ shown in fig. 16a, b. $|\Pi^W(0)|$ is lowered by $\simeq 0.011$. In figs. 16a, b we present also the variation of $\Pi^W(0)$ and $\Pi^{\gamma Z}(0)$ with the mass of the top quark. A change from $m_t = 30$ GeV to e.g. $m_t = 60$ GeV increases $|\Pi^W(0)|$ for $s_W^2 = 0.221$ by 0.002.

Finally we do not know the Higgs mass M_H. Therefore we have displayed $\Pi^W(0), \Pi^{\gamma Z}(0)$ also for $M_H = 10$ GeV and $M_H = 300$ GeV in figs. 16. A light Higgs mass decreases $|\Pi^W(0)|$, a heavy one increases it. The conclusion of this discussion is that our ignorance of the mass parameters give uncertainties in the calculation of Π^W, Π^Z and the other self energies amounting to ± 0.01 which might be of the same order of magnitude as 2-loop effects.

6.2 Fermion self energies

We described the renormalization prescription for the lepton and quark self energies in sect. 5.3. Together with the unrenormalized expressions (5.27) and the renormalization constants (5.43—46) we obtain the renormalized self energies from the equations:

$$\hat{\Sigma}^{i\sigma}(k) = \not{k}\,\frac{1-\gamma_5}{2}\left(\Sigma_L^{i\sigma}(k^2) + \delta Z_L^i\right) + \not{k}\,\frac{1+\gamma_5}{2}\left(\Sigma_R^{i\sigma}(k) + \delta Z_R^{i\sigma}\right)$$
$$+ m_{i\sigma}\Sigma_S^{i\sigma}(k^2) - m_{i\sigma}(\delta Z_R^{i\sigma} + \delta Z_L^i)/2 - \delta m_{i\sigma}. \tag{6.6}$$

We illustrate the results with help of the ν_e self energy and the electron self energy. The corresponding invariant functions are shown in figs. 17 and 18. For the neutrinos only left-handed contributions exist. They are in our renormalization scheme infrared divergent. Therefore in fig. 17 the IR finite quantity $\hat{\Sigma}_L^\nu(p^2) - \hat{\Sigma}_L^\nu(0)$ is drawn for time-like momenta p^2. It depends only weakly on p^2.

The real and imaginary parts of the invariant functions $\hat{\Sigma}_V^e$, $\hat{\Sigma}_A^e$, $\hat{\Sigma}_S^e$ of the electron self energy are presented in figs. 18. In the case of Re $\hat{\Sigma}_V^e$ and Re $\hat{\Sigma}_S^e$ we have subtracted the IR divergent part $\alpha/4\pi\left(2\ln(m_e^2/\lambda^2) - 4\right)$. We find that $\hat{\Sigma}_A$ and $\hat{\Sigma}_V$ are small, only Re $\hat{\Sigma}_S^e(p^2)$ reaches a level of several percent.

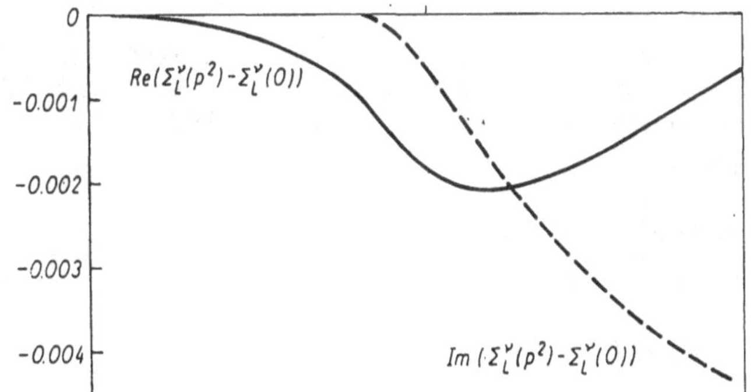

Fig. 17. Real (——) and imaginary (- - - -) parts of the neutrino self energy subtracted at $p^2 = 0$ (Read 40^2, 80^2, ... instead of 40, 80, ... resp.)

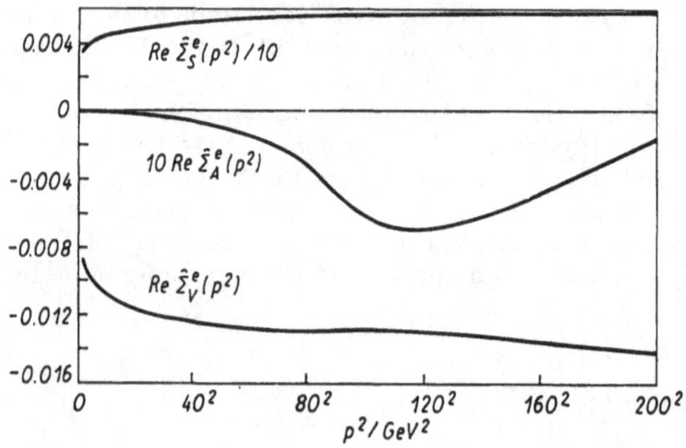

Fig. 18a

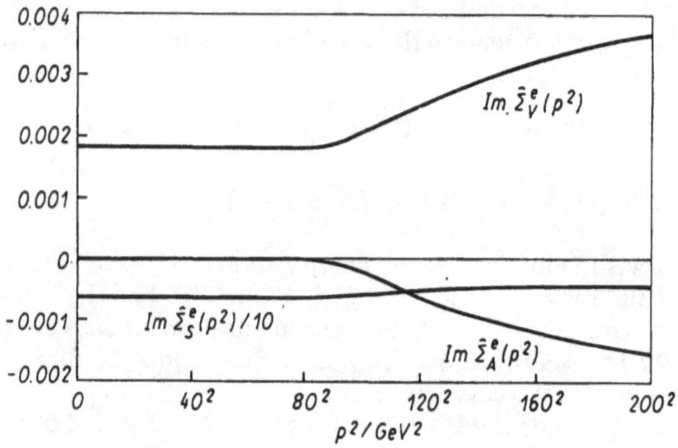

Fig. 18b

Fig. 18a, b. Real (a) and imaginary (b) parts of the electron self energy. Presented are the infrared finite parts of the invariant functions $\hat{\Sigma}_S{}^e$, $\hat{\Sigma}_A{}^e$, $\hat{\Sigma}_V{}^e$

6.3 Renormalized gauge boson fermion vertices

The following list of the renormalized vertex functions contains vector and axial vector couplings only and is valid for on shell fermions and $|k^2| \gg m_f{}^2$. As in sect. 5.2 we write down the formulas only for the first lepton and quark multiplet.

a) Electromagnetic current:

$$\begin{aligned}
\hat{\Gamma}_\mu{}^{\gamma ee}(k^2) &= ie\gamma_\mu(F_V{}^{\gamma e} - \gamma_5 F^{\gamma e}),\\
\hat{\Gamma}_\mu{}^{\gamma\nu\nu}(k^2) &= ie\gamma_\mu(1 - \gamma_5)\, F^{\gamma\nu},\\
\hat{\Gamma}_\mu{}^{\gamma dd}(k^2) &= -ieQ_d\gamma_\mu(F_V{}^{\gamma d} - \gamma_5 F_A{}^{\gamma d}),\\
\hat{\Gamma}_\mu{}^{\gamma uu}(k^2) &= -ieQ_u\gamma_\mu(F_V{}^{\gamma u} - \gamma_5 F^{\gamma u}).
\end{aligned} \tag{6.7}$$

The form factors contain the functions $\Lambda_{1,\dots,4}$ given in Appendix B.3:

$$F_V{}^{\gamma e} = 1 + \frac{\alpha}{4\pi}\left[\Lambda_1(k^2, m_e) + (v_e{}^2 + a_e{}^2)\, \Lambda_2(k^2, M_Z) + \frac{3}{4s^2}\, \Lambda_3(k^2, M_W)\right],$$

$$F_A^{\gamma e} = \frac{\alpha}{4\pi}\left[2v_e a_e \Lambda_2(k^2, M_Z) + \frac{3}{4s^2}\Lambda_3(k^2, M_W)\right], \tag{6.8}$$

$$F^{\gamma\nu} = \frac{\alpha}{4\pi}\frac{1}{4s^2}\left[\Lambda_2(k^2, M_W) - 3\Lambda_3(k^2, M_W)\right];$$

$$F_V^{\gamma d} = 1 + \frac{\alpha}{4\pi}\left[Q_d^2\Lambda_1(k^2, m_d) + (v_d^2 + a_d^2)\Lambda_2(k^2, M_Z) - \frac{1}{2s^2}\Lambda_2(k^2, M_W) + \frac{9}{4s^2}\Lambda_3(k^2, M_W)\right],$$

$$F_A^{\gamma d} = \frac{\alpha}{4\pi}\left[2v_d a_d \Lambda_2(k^2, M_Z) - \frac{1}{2s^2}\Lambda_2(k^2, M_W) + \frac{9}{4s^2}\Lambda_3(k^2, M_W)\right], \tag{6.9}$$

$$F_V^{\gamma u} = 1 + \frac{\alpha}{4\pi}\left[Q_u^2\Lambda_1(k^2, m_u) + (v_u^2 + a_u^2)\Lambda_2(k^2, M_Z) - \frac{1}{8s^2}\Lambda_2(k^2, M_W) + \frac{9}{8s^2}\Lambda_3(k^2, M_W) + \delta(u, d)\right]^{9)},$$

$$F_A^{\gamma u} = \frac{\alpha}{4\pi}\left[2v_u a_u \Lambda_2(k^2, M_Z) - \frac{1}{8s^2}\Lambda_2(k^2, M_W) + \frac{9}{8s^2}\Lambda_3(k^2, M_W)\right].$$

b) Weak neutral current:

$$\begin{aligned}
\hat{\Gamma}_\mu^{Zee}(k^2) &= ie\gamma_\mu(F_V^{Ze} - \gamma_5 F_A^{Ze}),\\
\hat{\Gamma}_\mu^{Z\nu\nu}(k^2) &= ie\gamma_\mu(1 - \gamma_5)\, F^{Z\nu},\\
\hat{\Gamma}_\mu^{Zdd}(k^2) &= ie\gamma_\mu(F_V^{Zd} - \gamma_5 F_A^{Zd}),\\
\hat{\Gamma}_\mu^{Zuu}(k^2) &= ie\gamma_\mu(F_V^{Zu} - \gamma_5 F_A^{Zu})
\end{aligned} \tag{6.10}$$

with the form factors:

$$F_V^{Ze} = v_e + \frac{\alpha}{4\pi}\left[v_e\Lambda_1(k^2, m_e) + v_e(v_e^2 + 3a_e^2)\Lambda_2(k^2, M_Z) + \frac{1}{8s^3c}\Lambda_2(k^2, M_W) - \frac{3c}{4s^3}\Lambda_3(k^2, M_W)\right],$$

$$F_A^{Ze} = a_e + \frac{\alpha}{4\pi}\left[a_e\Lambda_1(k^2, m_e) + a_e(3v_e^2 + a_e^2)\Lambda_2(k^2, M_Z) + \frac{1}{8s^3c}\Lambda_2(k^2, M_W) - \frac{3c}{4s^3}\Lambda_3(k^2, M_W)\right], \tag{6.11}$$

$$F^{Z\nu} = \frac{1}{4sc}\left\{1 + \frac{\alpha}{4\pi}\left[-\ln\frac{M_Z^2}{m_e^2} - \frac{9}{2} + 2\ln\frac{m_e^2}{\lambda^2} + \frac{1}{4s^2c^2}\Lambda_2(k^2, M_Z) + \frac{2s^2-1}{2s^2}\Lambda_2(k^2, M_W) + \frac{3c^2}{s^2}\Lambda_3(k^2, M_W)\right]\right\};$$

9) $\delta(u, d)$ is defined in eq. (5.46).

$$F_V^{Zd} = v_d + \frac{\alpha}{4\pi}\left[v_d Q_d^2 \Lambda_1(k^2, m_d) + v_d(v_d^2 + 3a_d^2)\,\Lambda_2(k^2, M_Z) + \frac{1 - 2Q_u s^2}{8s^3 c}\,\Lambda_2(k^2, M_W) - \frac{3c}{4s^3}\,\Lambda_3(k^2, M_W)\right],$$

$$F_A^{Zd} = a_d + \frac{\alpha}{4\pi}\left[a_d Q_d^2 \Lambda_1(k^2, m_d) + a_d(3v_d^2 + a_d^2)\,\Lambda_2(k^2, M_Z) + \frac{1 - 2Q_u s^2}{8s^3 c}\,\Lambda_2(k^2, M_W) - \frac{3c}{4s^3}\,\Lambda_3(k^2, M_W)\right],$$

$$F_V^{Zu} = v_u + \frac{\alpha}{4\pi}\left[v_u Q_u^2 \Lambda_1(k^2, m_u) + v_u(v_u^2 + 3a_u^2)\,\Lambda_2(k^2, M_Z) + v_u\delta(u, d) - \frac{1 + 2Q_d s^2}{8s^3 c}\,\Lambda_2(k^2, M_W) + \frac{3c}{4s^3}\,\Lambda_3(k^2, M_W)\right],$$

$$F_A^{Zu} = a_u + \frac{\alpha}{4\pi}\left[a_u Q_u^2 \Lambda_1(k^2, m_u) + a_u(3v_u^2 + a_u^2)\,\Lambda_2(k^2, M_Z) + a_u\delta(u, d) - \frac{1 + 2Q_d s^2}{8s^3 c}\,\Lambda_2(k^2, M_W) + \frac{3c}{4s^3}\,\Lambda_3(k^2, M_W)\right]. \tag{6.12}$$

c) Weak charged current:

$$\hat{\Gamma}_\mu^{We\nu}(k^2) = i\,\frac{e}{2\sqrt{2}\,s}\,\gamma_\mu(1 - \gamma_5)\,F^{We\nu},$$

$$\hat{\Gamma}_\mu^{Wdu}(k^2) = i\,\frac{e}{2\sqrt{2}\,s}\,\gamma_\mu(1 - \gamma_5)\,F^{Wdu}, \tag{6.13}$$

with the formfactors

$$F^{We\nu} = 1 + \frac{\alpha}{4\pi}\left\{\frac{3(3c^2 - 1)}{2s^2} + \ln\frac{m_e^2}{M_W^2} + 2\ln\frac{m_e^2}{\lambda^2} + \left(\frac{2s^2 - 1}{2s^2} + \frac{3c^2}{s^4}\right)\ln\frac{M_W^2}{M_Z^2} + \frac{2s^2 - 1}{4s^2c^2}\,\Lambda_2(k^2, M_Z) + 3\Lambda_4^{m_e}(k^2; M_W, 0) + \frac{3c^2}{s^2}\,\Lambda_4(k^2; M_Z, M_W)\right\},$$

$$\begin{aligned} F^{Wdu} = 1 + \frac{\alpha}{4\pi}\Bigg\{&Q_u Q_d\left[3\ln\frac{m_d}{m_u} + \frac{1}{2}\,\Lambda_1(k^2, m_d) + \frac{1}{2}\,\Lambda_1(k^2, m_u) + \frac{s^2}{c^2}\,\Lambda_2(k^2, M_Z)\right] + \frac{2s^2 - 1}{4s^2c^2}\,\Lambda_2(k^2, M_Z) + 3Q_u\Lambda_4^{m_u}(k^2, M_W, 0) \\ &- 3Q_d\Lambda_4^{m_d}(k^2, M_W, 0) + \frac{3c^2}{s^2}\,\Lambda_4(k^2, M_Z, M_W) \\ &- \frac{1}{3}\left(\ln\frac{M_Z^2}{m_d^2} - 2\ln\frac{m_d^2}{\lambda^2} + \frac{9}{2}\right) + \frac{3}{s^2} + \left(\frac{1}{2s^2} - \frac{3c^2}{s^4}\right)\ln\frac{M_Z^2}{M_W^2}\Bigg\}. \end{aligned} \tag{6.14}$$

d) Examples: The electron and ν photon formfactors, the electron Z boson formfactor, the $We\nu$-formfactor

For illustration we present the weak contributions (the parts with $\Lambda_{2,3}$ in eq.s (6.7—13)) to the vector and axialvector photonic formfactor of the electron in fig. 19 for $|k^2| \leqq (150\text{ GeV})^2$. The vector part $F^{\gamma e}_{V,\text{weak}}$ vanishes at $k^2 = 0$ as a consequence of charge renormalization, the axial vector part $F^{\gamma e}_{A,\text{weak}}(0) = 0$ because of the Ward identity. For the k^2 values given above the real and imaginary parts of these formfactors are typically of the order of magnitude if $10^{-3}e$.

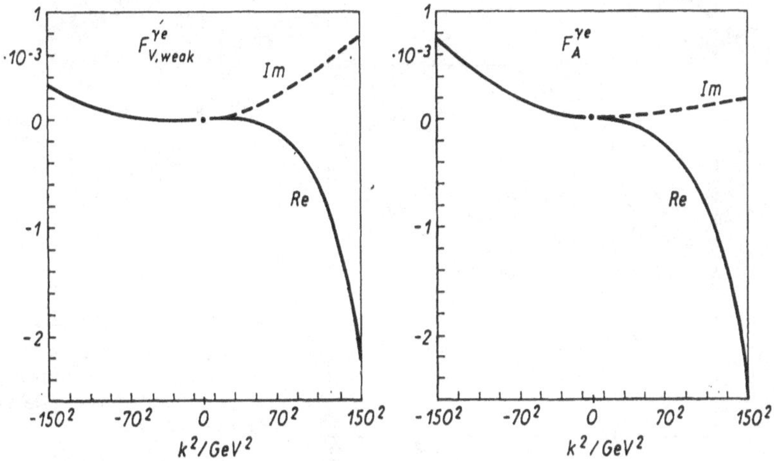

Fig. 19a, b. Real (——) and imaginary (- - - -) parts of the weak corrections to the vector (a) and axial vector (b) photon formfactor of the electron

The $\gamma\nu$ vertex vanishes in lowest order but gets contributions from 1-loop diagrams (b, c of fig. 7) containing the W exchange and the non-Abelian gauge boson coupling. The resulting formfactor $F^{\gamma\nu}(k^2)$ shown in fig. 20 grows for $|k^2| \leqq (150\text{ GeV})^2$ to $\sim 10^{-3}e$.

The non-photonic contributions to the eZ boson formfactors are shown in fig. 21a, b. They have the same characteristics as the other formfactors. Compared to the self energy

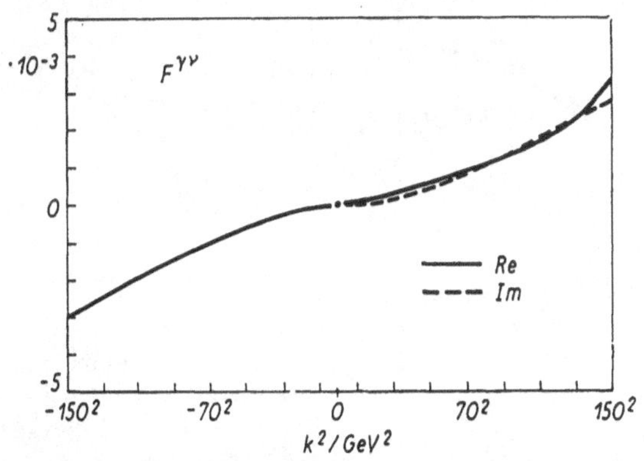

Fig. 20. Real (——) and imaginary (- - - -) parts of the electromagnetic formfactor of the neutrino

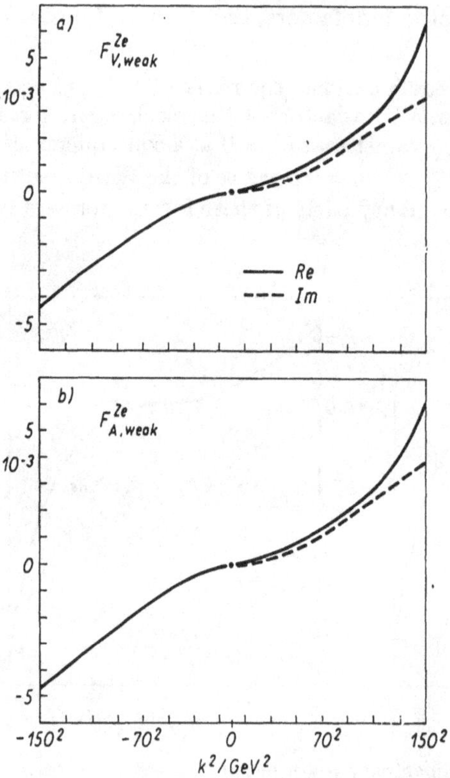

Fig. 21 a, b. Real (——) and imaginary (– – – –) parts of the weak contributions to the vector (a) and axial vector (b) Z boson electron formfactor

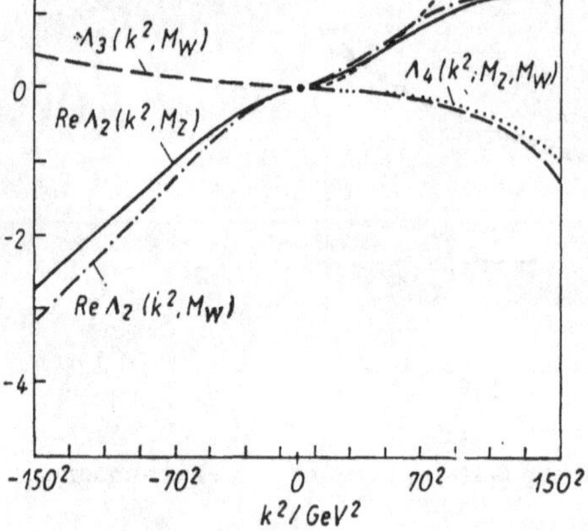

Fig. 22. Real and imaginary parts of the vertex integral $\Lambda_2(k^2, M)$ for $M = M_W$, M_Z and $\Lambda_3(k^2, M_W)$, $\Lambda_4(k^2, M_Z, M_W)$

effects of the weak bosons the weak contributions to the vertex corrections give effects in e^+e^- annihilation processes which are one order of magnitude smaller.

The $We\nu$- and the corresponding quark formfactors obtain a contribution from the $WW\gamma$-coupling (see fig. 7c). This shows a pronounced structure around $k^2 = M_W^2$ and reaches an order of magnitude of $-40 \cdot \alpha/4\pi$ for the real and imaginary part (see fig. 23).

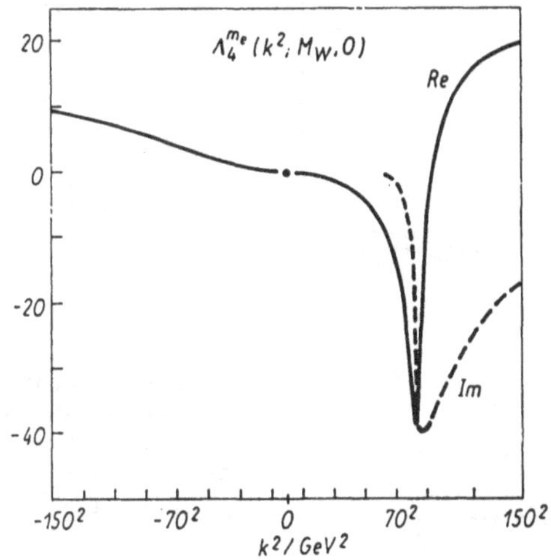

Fig. 23. Real (——) and imaginary (- - - -) parts of the photonic invariant function $\Lambda_4^m(k^2, M_W, 0)$ for $m = m_e$

Together with the corresponding contribution (the $W\gamma$-loop) to the W self energy it may lead to interesting effects in W exchange dominated reactions for timelike large momentum transfers.

We conclude this section with some remarks on the box diagrams with two weak bosons. In contrast to the self energy and vertex diagrams they are both UV and IR finite and consequently in the 1-loop approximation not directly influenced by the renormalization scheme. Their contribution to 1-loop radiative corrections to S matrix elements is in the energy range considered of the same order of magnitude as those of the vertex corrections. Explicit expressions are given in the appendix B.4 and numerical results presented in fig. 24a, b.

7. Application to Purely Leptonic Reactions

In this section we apply these 1-loop results to purely leptonic reactions like μ decay, $\overset{(-)}{\nu}_\mu$e scattering and lepton pair production in e^+e^- annihilation [29]. Although the results from deep inelastic lepton scattering have now reached very good statistics, we have restricted our analysis to leptonic processes in this paper since these have smaller theoretical uncertainties from the strong interaction. Therefore these processes allow the cleanest tests of the electroweak interaction.

According to the choice of M_W, M_Z as parameters in the renormalization scheme used, the most direct way to compare the predictions of the electroweak standard theory with experimental data is to start with the measured values for M_W, M_Z (or equivalently M_W, s_W^2 resp. M_Z, s_W^2).

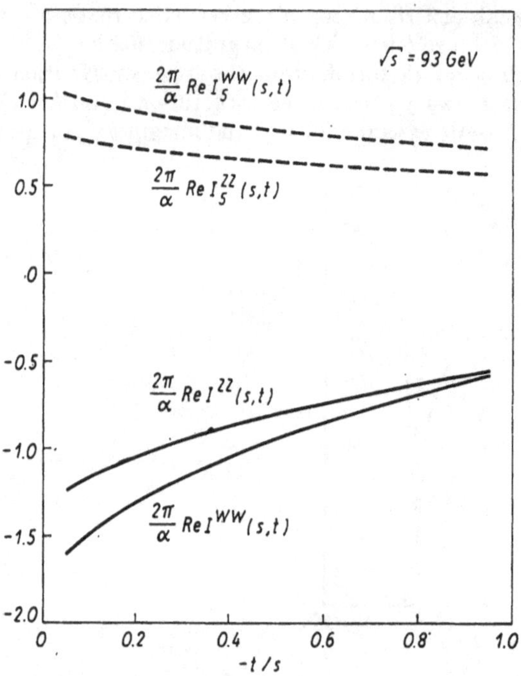

Fig. 24a. Real part of the s-channel box diagram form-factors $2\pi/\alpha \cdot I$ (——) and $2\pi/\alpha \cdot I_5$ (– – – –) for the ZZ and WW box $\left(\sqrt{s} = 93\ \text{GeV}\right)$

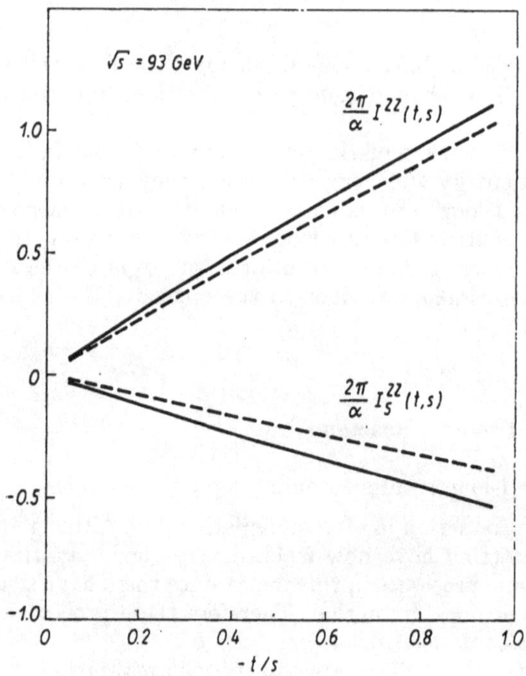

Fig. 24b. Real (——) and imaginary (– – – –) parts of the t-channel box $2\pi/\alpha \cdot I^{ZZ}(t, s)$ and $2\pi/\alpha \cdot I_5{}^{ZZ}(t, s)$ with two Z bosons for $\sqrt{s} = 93$ GeV

The UA 1 and UA 2 groups have determined

$$M_W = 82.2 \pm 1.1 \text{ GeV}, \qquad M_Z = 92.7 \pm 1.0 \text{ GeV}. \tag{7.1}$$

A value for $\Delta M = M_Z - M_W$ can be deduced from experiment with an error smaller than that resulting from (7.1) because of a partial cancellation of the systematic uncertainties. Together with the definition of s_W^2:

$$s_W^2 = 1 - \frac{M_W^2}{M_Z^2} = \frac{\Delta M}{M_Z}\left(2 - \frac{\Delta M}{M_Z}\right) \tag{7.2}$$

this gives for the mixing angle [1]:

$$s_W^2 = 0.218 \pm 0.023. \tag{7.3}$$

a) μ decay:

Neglecting terms of order $(m_e/m_\mu)^3$ and $(m_\mu/M_W)^2$ the lowest order expression for the decay width for $\mu^- \to \nu_\mu \bar{\nu}_e e^-$ is given by:

$$\begin{aligned} \Gamma_\mu^0 &= \frac{\alpha^2}{384\pi} m_\mu \left(1 - 8\frac{m_e^2}{m_\mu^2}\right) \frac{m_\mu^4}{M_W^4} \frac{1}{(1 - M_W^2/M_Z^2)^2} \\ &= \frac{\alpha^2}{384\pi} m_\mu \left(1 - 8\frac{m_e^2}{m_\mu^2}\right) \left(\frac{m_\mu}{M_W s_W}\right)^4. \end{aligned} \tag{7.4}$$

Putting together the 1-loop corrections yields the following results:

$$\begin{aligned} \Gamma_\mu &= \Gamma_\mu^0 \left\{1 + \frac{\alpha}{2\pi}\left(\frac{25}{4} - \pi^2\right) - 2\Pi^W(0) + \frac{\alpha}{2\pi s_W^2}\left[6 + \frac{7 - 4s_W^2}{2s_W^2} \ln c_W^2\right]\right\} \\ &= \Gamma_\mu^0(1 + \delta\Gamma_\mu/\Gamma_\mu^0). \end{aligned} \tag{7.5}$$

The first correction term is the familiar QED correction in the Fermi model, the second the contribution of the transverse part of the W self energy, $\Pi^W(0)$ of eq. (6.4), the last term the sum of the vertex and box diagrams together with the ν_e, ν_μ wave function renormalization.

The decay width $\Gamma_\mu = \Gamma_\mu^0 + \delta\Gamma_\mu$ depends on both M_W and s_W^2. Using the mean values of (7.1) and (7.3) the expressions (7.4–5) give:

$$\Gamma_\mu = 2.96 \cdot 10^{-16} \text{ MeV},$$

whereas the QED corrected Fermi model result is:

$$\Gamma_\mu^0 \left[1 + \frac{\alpha}{2\pi}\left(\frac{25}{4} - \pi^2\right)\right] = 2.56 \cdot 10^{-16} \text{ MeV}.$$

This has to be compared with the measured value $\Gamma_\mu^{exp} = 2.9958 \cdot 10^{-16}$ MeV [26]. With a fixed value of $M_W = 82.2$ GeV we obtain $2.42 \cdot 10^{-16} \leqq \Gamma_\mu \leqq 3.70 \cdot 10^{-16}$ MeV, corresponding to the variation of s_W^2 in (7.3).

The present accuracy of the direct M_W, M_Z measurements does not allow to predict Γ_μ with a precision that can compete with the accuracy of Γ_μ^{exp}. Instead, Γ_μ^{exp} can be used as an input quantity from which for a given s_W^2 (resp. M_W) the corresponding M_W and M_Z (resp. s_W^2, M_Z) is obtained. The result of this calculation is shown in figs. 25 and 26 for various values of the other parameters.

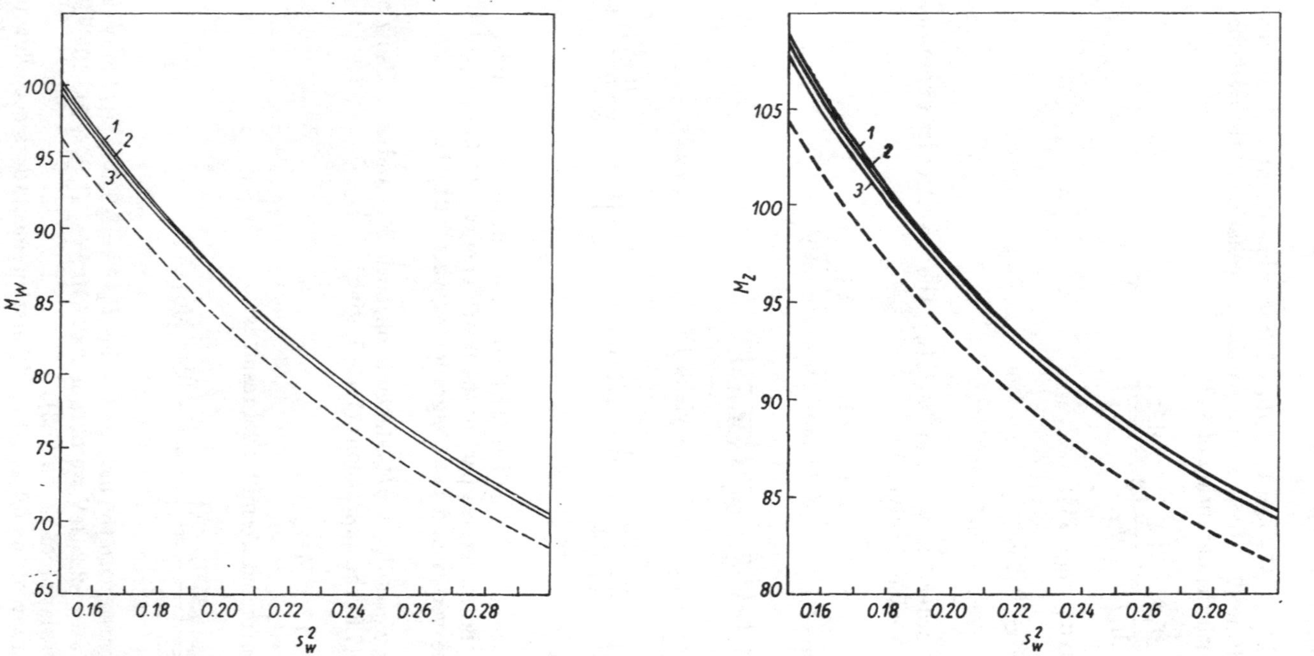

Fig. 25. M_W (a) and M_Z (b) as function of $s_W{}^2$ determined from the μ decay width in lowest order (– – – –) and including 1-loop radiative corrections (———) for:

1: $(m_u, m_d, m_s) = (5, 7, 150)$ MeV, $m_t = 60$ GeV;
2: $(m_u, m_d, m_s) = (5, 7, 150)$ MeV, $m_t = 30$ GeV;
3: $(m_u, m_d, m_e) = (300, 300, 450)$ MeV, $m_t = 30$ GeV;
$M_H = 100$ GeV

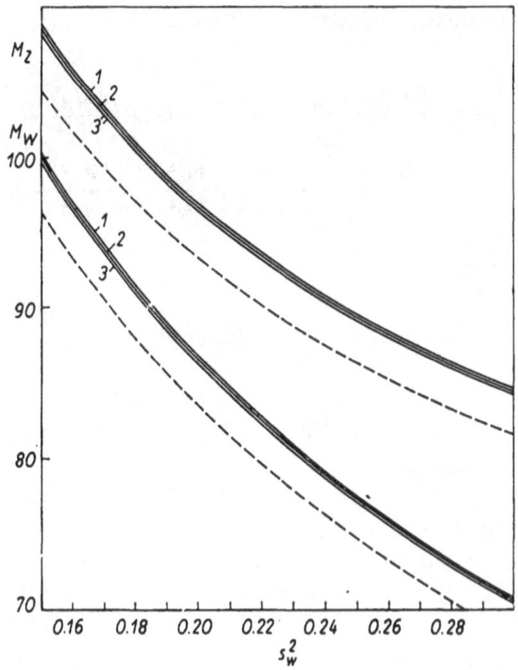

Fig. 26. Same as fig.s 25a, b for several choices of M_H: 1: $M_H = 300$ GeV, 2: $M_H = 100$ GeV, 3: $M_H = 10$ GeV. Quark masses as in eq. (6.2)

b) $\overset{(-)}{\nu}_\mu e$ scattering:

The ratio

$$\Delta_\nu = \frac{\sigma(\nu_\mu e) - \sigma(\bar{\nu}_\mu e)}{\sigma(\nu_\mu e) + \sigma(\bar{\nu}_\mu e)} = \frac{\sigma(\nu_\mu e)/\sigma(\bar{\nu}_\mu e) - 1}{\sigma(\nu_\mu e)/\sigma(\bar{\nu}_\mu e) + 1} = \frac{R_\nu - 1}{R_\nu + 1} \tag{7.6}$$

is well-suited for our purpose since it is sensitive to the ratio M_W/M_Z resp. $s_W{}^2$ and less subject to systematic errors than the cross sections themselves; moreover it is free of electromagnetic higher order corrections.

With the ratio ξ of the vector and axial vector coupling constants of the electron to the Z:[10])

$$\xi = v/a = 1 - 4s_W{}^2. \tag{7.7}$$

Δ_ν has in lowest order the simple form:

$$\Delta_\nu{}^0 = \frac{\xi}{1 + \xi^2} \quad \text{or} \quad R_\nu{}^0 = \frac{1 + \xi + \xi^2}{1 - \xi + \xi^2}. \tag{7.8}$$

The 1-loop corrected expression of Δ_ν can be written in the following way:

$$\Delta_\nu = \frac{\xi + \Delta^{\gamma Z} - V + \xi A}{1 + \xi^2 + A\xi^2 + 2\xi(\Delta^{\gamma Z} - V)} = \Delta_\nu{}^0 + \delta\Delta_\nu. \tag{7.9}$$

[10]) In this section we omit the index e of the coupling constants v_e, a_e.

It gets contributions from the γZ mixing energy $\hat{\Sigma}_T^{\gamma Z}$ (6.1), the derivative of the $\gamma\nu\nu$ formfactor

$$\Delta^{\gamma Z} = 4c_W s_W \frac{1}{s} \hat{\Sigma}_T^{\gamma Z}(s)|_{s=0} + \frac{2\alpha}{3\pi}\left(\ln\frac{M_W^2}{m_\mu^2} + 1\right) = -4c_W^2\left(\frac{\delta M_Z^2}{M_Z^2} - \frac{\delta M_W^2}{M_W^2}\right)\Bigg|_{\text{fin}}$$
$$+ \frac{\alpha}{3\pi}\left(3 + 2c_W^2 + 2\ln\frac{M_W^2}{m_\mu^2}\right) - \frac{\alpha}{4\pi}\frac{2}{3}\left(\ln\frac{m_u^2}{m_d^2} + \ln\frac{m_c^2}{m_s^2} + \ln\frac{m_t^2}{m_b^2}\right)$$
$$- \frac{\alpha}{4\pi}\frac{c^2}{s^2}\frac{m_t^2}{M_W^2}\ln\frac{m_t^2}{m_b^2} \tag{7.10}$$

and of the box diagrams containing two massive gauge bosons:

$$V - +\frac{\alpha}{\pi}\left[\frac{1}{s_W^2} + 3va\right], \qquad A = -\frac{\alpha}{\pi}\left[\frac{1}{s_W^2} + \frac{3}{2}(v^2 + a^2)\right]. \tag{7.11}$$

The weak contributions to the renormalized $\gamma\nu\nu$, $Z\nu\nu$ and Zee vertex functions vanish in our scheme at zero momentum transfer, yielding the simple expressions above.

The quantity R_ν resp. Δ_ν, eq. (7.6), depends on M_W and M_Z mainly via the combination M_W/M_Z, because the variation of the γZ mixing energy with M_W (whence s_W^2 fixed) is small (see fig. 15). Therefore the value of s_W^2 from (7.3) can directly be converted into the observable R_ν (see also fig. 27):

$$R_\nu^0 = 1.29^{+0.24}_{-0.21}, \qquad R_\nu = 1.28^{+0.23}_{-0.20}.$$

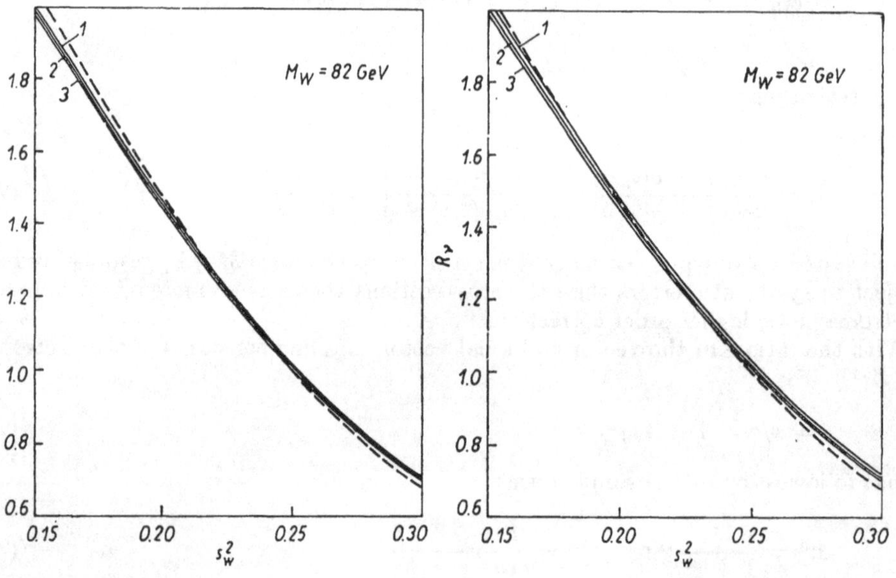

Fig. 27. R_ν as function of s_W^2 in lowest order (- - - -) and including radiative corrections (——) for several choices of

a) M_H; 1: $M_H = 300$ GeV, 2: $M_H = 100$ GeV, 3: $M_H = 10$ GeV and
b) quark masses:

1: $(m_u, m_d, m_s) = (300, 300, 450)$ MeV, $m_t = 30$ GeV;
2: $(m_u, m_d, m_s) = (5, 7, 150)$ MeV, $m_t = 60$ GeV;
3: $(m_u, m_d, m_s) = (5, 7, 150)$ MeV, $m_t = 30$ GeV; other parameters from (6.2)

The actual experimental value is [30]:

$$R_\nu = 1.38^{-0.31}_{+0.40}.$$

The measurement of R_ν together with that of Γ_μ an be used to determine the gauge boson masses by simultaneously solving the equations

$$R_\nu^{\text{exp}} = R_\nu^{\text{theor}}(M_W, M_Z), \qquad \Gamma_\mu^{\text{exp}} = \Gamma_\mu^{\text{theor}}(M_W, M_Z).$$

This way to analyse the low energy data has been — until the experimental discovery of the W and Z — the only possibility to get information on the values of the gauge boson masses from purely leptonic reactions [31]. The results of our calculation are presented in fig. 28, where we have plotted the W and Z mass as a function of R_ν^{exp}.

c) Forward-backward asymmetry in $e^+e^- \to \mu^+\mu^-$:

The forward backward asymmetry $A_{FB}(x)$ in e^+e^- annihilation into μ pairs is defined as ($c = \cos\theta$):

$$A_{FB}(x) = \frac{\int\limits_0^x dc \frac{d\sigma}{d\Omega} - \int\limits_{-x}^0 dc \frac{d\sigma}{d\Omega}}{\int\limits_0^x dc \frac{d\sigma}{d\Omega} + \int\limits_{-x}^0 dc \frac{d\sigma}{d\Omega}}. \tag{7.12}$$

$d\sigma/d\Omega$ reads in Born approximation

$$\frac{4s}{\alpha^2} \cdot \frac{d\sigma^0}{d\Omega} = 1 + c^2 + 2\chi(s) \left[v^2(1+c)^2 + 2a^2c\right] + \chi(s)^2 \left[(v^2+a^2)^2 (1+c^2) + 4v^2a^2 \cdot 2c\right] \tag{7.13}$$

where:

$$\chi(s) = \frac{s}{s - M_Z^2}. \tag{7.14}$$

This gives:

$$A^0_{FB}(x, s) = \frac{x}{1 + x^2/3} \cdot 2a^2\chi(s) \frac{1 + 2v^2\chi(s)}{1 + 2v^2\chi(s) + (v^2+a^2)^2 \chi(s)^2}. \tag{7.15}$$

At PETRA/PEP energies one is allowed to neglect the imaginary part $M_Z\Gamma_Z$ in the denominator of χ since $(\Gamma_Z/M_Z)^2 \ll 1$. Also we know that $v^2 \ll a^2$ and consequently may simplify (7.15):

$$A_{FB} = A^0_{FB}(1, s) \cong \frac{3}{2} a^2\chi/(1 + a^4\chi^2) \tag{7.16}$$

with

$$a^2\chi = -\frac{s}{16M_W^2(1 - M_W^2/M_Z^2)} \frac{M_Z^2}{M_Z^2 - s}. \tag{7.17}$$

The radiative corrections to $d\sigma/d\Omega$ can be divided into electromagnetic (real and virtual photonic corrections) and purely weak parts:

$$\frac{d\sigma}{d\Omega} = \frac{d\sigma^0}{d\Omega} (1 + C_{\text{em}} + C_W). \tag{7.18}$$

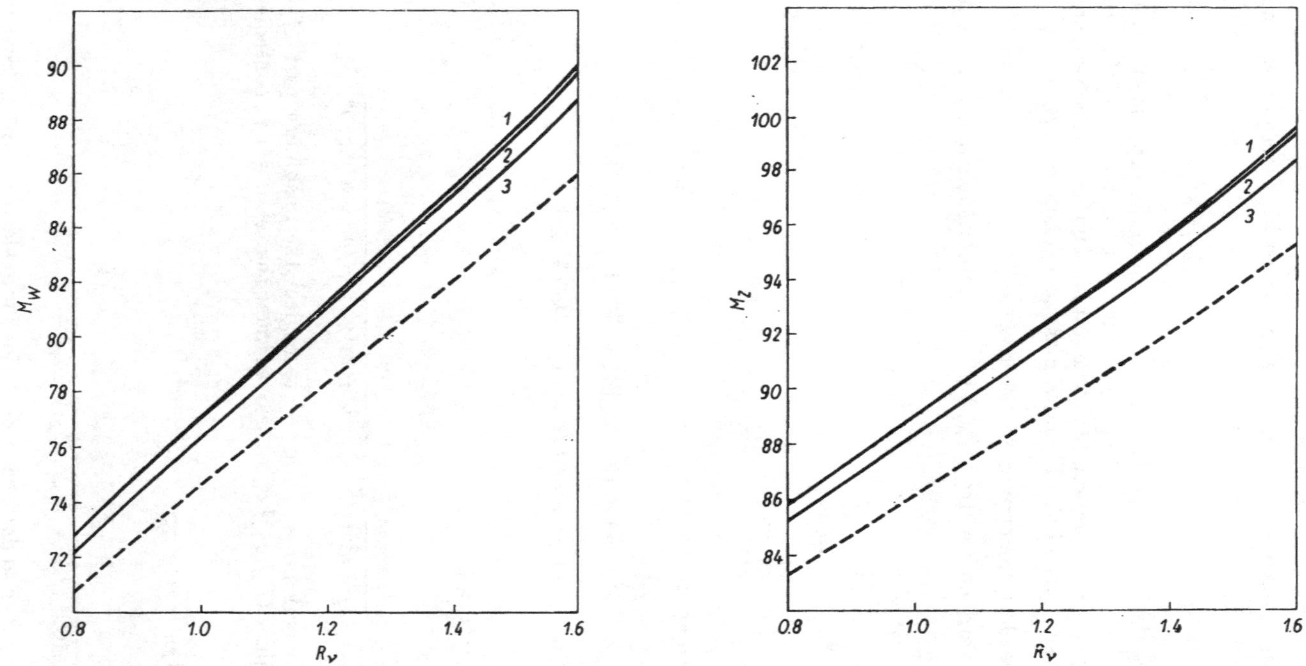

Fig. 28a, b. M_W (a) and M_Z (b) determined from μ decay and $\nu_\mu e$ scattering as function of R_ν in lowest order (- - - -) and including radiative corrections (——) for:

1: $(m_u, m_d, m_s) = (5, 7, 150)$ MeV, $m_t = 30$ GeV;
2: $(m_u, m_d, m_s) = (5, 7, 150)$ MeV, $m_t = 60$ GeV;
3: $(m_u, m_d, m_s) = (300, 300, 450)$ MeV, $m_t = 30$ GeV;
$M_H = 100$ GeV

The electromagnetic corrections C_{em} and their influence on $\mathcal{A}_{FB}$ have been treated in ref. [18] and especially in [19]. Therefore we do not reproduce the expressions for C_{em} in this paper but take the formulas of [19] for the numerical evaluation of their contribution to $\mathcal{A}_{FB}$.

The purely weak part C_w is built up from the Z self energy, the gauge part of the γ self energy, the γZ mixing energy, the weak contributions F_w to the e and μ photon and Z formfactors and the box graphs with two heavy bosons. For PETRA/PEP energies, neglecting terms of order $\alpha/2\pi \cdot (|t|/M_Z^2)$ these box contributions become independent of $c = \cos\theta$, and therefore the weak corrections can be written in the following way:

$$\frac{d\sigma^0}{d\Omega} \cdot C_w = (1 + c^2) \left[C_w^{\gamma,+} + 2\chi C_w^{\gamma Z,+} + \chi^2 C_w^{Z,+}\right] + 2c\left[2\chi C_w^{\gamma Z,-} + \chi^2 C_w^{Z,-}\right]. \tag{7.19}$$

These terms modify the expression (7.16) for the forward-backward asymmetry $\mathcal{A}_{FB}$ to become:

$$\mathcal{A}_{FB} = \frac{3}{4} \cdot \frac{2\chi(a^2 + C_w^{\gamma Z,-}) + \chi^2(4v^2a^2 + C_w^{Z,-})}{1 + C_w^{\gamma,+} + 2\chi(v^2 + C_w^{\gamma Z,+}) + \chi^2\left((v^2 + a^2)^2 + C_w^{Z,+}\right)}. \tag{7.20}$$

Now we write down the explicit form of the corrections C_w:

$$\begin{aligned}
C_w^{\gamma,+} &= -2\Pi_w^{\gamma} + 4F^{\gamma,e}_{V,w},\\
C_w^{\gamma Z,+} &= -v^2(\Pi_w^{\gamma} + \Pi^{\gamma Z}) - 2v\Pi^{\gamma Z} + 2vF^{Ze}_{V,w} + 2v(vF^{\gamma e}_{V,w} + aF^{\gamma e}_{A,w})\\
&\quad + 4v^2a^2A_1^{ZZ} + (2s_W)^{-4}\,V_1^{WW},\\
C_w^{\gamma Z,-} &= -a^2(\Pi_w^{\gamma} + \Pi^{Z}) + 2aF^{Ze}_{A,w} + 2a(vF^{\gamma e}_{A,w} + aF^{\gamma e}_{V,w})\\
&\quad + (v^2 + a^2)^2\,A_1^{ZZ} + (2s_W)^{-4}\,V_1^{WW},\\
C_w^{Z,+} &= -2(v^2 + a^2)^2\,\Pi^Z - 2v(v^2 + a^2) \cdot 2\Pi^{\gamma Z} + 4(v^2 + a^2)\,(vF^{Ze}_{V,w} + aF^{Ze}_{A,w})\\
&\quad + \left(v \cdot 2va + a(v^2 + a^2)\right)^2 A_1^{ZZ} + (v + a)^2\,(2s_W)^{-4}\,V_1^{WW},\\
C_w^{Z,-} &= -8v^2a^2\Pi^Z - 8a^2v\Pi^{\gamma Z} + 8va(vF^{Ze}_{A,w} + aF^{Ze}_{V,w})\\
&\quad + \left(v(v^2 + a^2) + a \cdot 2va\right)^2 A_1^{ZZ} + (v + a)^2\,(2s_W)^{-4}\,V_1^{WW}.
\end{aligned} \tag{7.21}$$

The quantities Π are related to the renormalized transverse self energies $\hat{\Sigma}_T(s)$ (6.1):

$$\begin{aligned}
\Pi_w^{\gamma}(s) &= \frac{1}{s}\,\mathrm{Re}\,\hat{\Sigma}^{\gamma}_{T,w}(s) \text{ (non-fermionic part)},\\
\Pi^{\gamma Z}(s) &= \frac{1}{s}\,\mathrm{Re}\,\hat{\Sigma}_T^{\gamma Z}(s),\\
\Pi^{Z,W}(s) &= \frac{1}{s - M^2_{Z,W}}\,\mathrm{Re}\,\hat{\Sigma}_T^{Z,W}(s).
\end{aligned} \tag{7.22}$$

The weak contributions $F^{Ze}_{V,w}, \ldots, F^{\gamma e}_{A,w}$ to the formfactors are built from the functions $\Lambda_{2,3}(s, M^2)$ and coupling constants (comp. 6.8, 11) and app. B.3):

$$\begin{aligned}
F^{Ze}_{V,w}(s) &= \frac{\alpha}{4\pi}\left[v(v^2 + a^2)\,\mathrm{Re}\,\Lambda_2(s, M_Z^2)\right.\\
&\quad \left. + \frac{1}{8s_W^3c_W}\,\mathrm{Re}\,\Lambda_2(s, M_W^2) - \frac{3c_W}{4s_W^3}\,\Lambda_3(s, M_W^2)\right],
\end{aligned} \tag{7.23}$$

$$F^{Ze}_{A,w}(s) = \frac{\alpha}{4\pi}\left[a(3v^2+a^2)\,\mathrm{Re}\,\Lambda_2(s,M_Z^2) + \frac{1}{8s_W{}^3c_W}\,\mathrm{Re}\,\Lambda_2(s,M_W^2) - \frac{3c_W}{4s_W{}^3}\,\Lambda_2(s,M_W^2)\right],$$

$$F^{\gamma e}_{V,w}(s) = \frac{\alpha}{4\pi}\left[(v^2+a^2)\,\mathrm{Re}\,\Lambda_2(s,M_Z^2) + \frac{3}{4s_W{}^2}\,\Lambda_3(s,M_W^2)\right], \tag{7.23}$$

$$F^{\gamma e}_{A,w}(s) = \frac{\alpha}{4\pi}\left[2va\cdot\mathrm{Re}\,\Lambda_2(s,M_Z^2) + \frac{3}{4s_W{}^2}\,\Lambda_3(s,M_W^2)\right].$$

These expressions vanish for $s = 0$ and are for energies $\sqrt{s} < 45$ GeV smaller than 10^{-3}. Finally the low energy approximations of the ZZ, WW box diagrams have the simple form (coupling constants removed):

$$A_1{}^{ZZ} = -3\,\frac{\alpha}{4\pi}, \qquad V_1{}^{WW} = \frac{\alpha}{4\pi} \tag{7.24}$$

yielding terms of the order of magnitude of less than 10^{-3}.

If an accuracy of the relative corrections to $\mathcal{A}_{FB}$ at PETRA/PEP energies of 10^{-3} is desired, one is allowed to neglect in the contributions to C_w all terms but the self energies. Then one gets for $\mathcal{A}_{FB}^{\text{Born+weak}}$ the following expression (using also $v^2 \ll a^2$):

$$\mathcal{A}_{FB}^{\text{Born+weak}} \simeq \frac{3}{2}\,\chi a^2\cdot\frac{1-\Pi_w{}^\gamma-\Pi^Z}{1-2\Pi_w{}^\gamma+\chi^2a^4} \simeq \frac{3}{2}\,\frac{\chi a^2}{1+\chi^2a^4}\,(1+\Pi_w{}^\gamma-\Pi^Z). \tag{7.25}$$

χ was defined in eq. (7.14) as the ratio of the free Z and γ propagators. Therefore the result (7.25) has the simple interpretation:

$$\mathcal{A}_{FB}^{\text{Born+weak}}(s) = \frac{3}{2}\,\frac{\chi(s)^{\text{Born+weak}}}{1+\chi(s)^2\,a^4}\,a^2, \qquad \chi^{\text{Born+weak}} = \frac{s+\hat{\Sigma}^\gamma_{T,w}(s)}{s-M_Z{}^2+\hat{\Sigma}_T{}^Z(s)}. \tag{7.26}$$

The lowest order expression for $\chi(s)$ has to be replaced by the renormalized one, the radiative corrections to a^2 can be neglected.

The weak i.e. non-Abelian gauge contribution to Π^γ, comes from vacuum polarization by W pairs, the corresponding ghosts and unphysical charged Higgses and results for small energies as can be seen from (6.1), (5.7) in:

$$\Pi_W{}^\gamma(s) = -\frac{\alpha}{4\pi}\cdot\frac{s}{2M_W{}^2} + O\left(\frac{\alpha}{4\pi}\left(\frac{s}{M_W{}^2}\right)^2\right). \tag{7.27}$$

Consequently this can also be neglected at the desired accuracy at PETRA/PEP energies, leaving the simple result suited for the practical calculations:

$$\mathcal{A}_{FB}^{\text{Born+weak}} = \frac{3}{2}\,\frac{\chi(s)\,a^2}{1+\chi^2a^4}\,\big(1-\Pi^Z(s)\big). \tag{7.28}$$

We find — in agreement with [16] — that the weak radiative corrections to $\mathcal{A}_{FB}$ at low energies are determined by the transverse Z boson self energy only, with an accuracy of $\delta\mathcal{A}^{\text{weak}}/\mathcal{A}^0 < 10^{-3}$.

On the basis of the formulas (7.18—23) we present in table 1 the weak corrections to A_{FB} for a large range of the Z mass and s_W^2. The α^3 contribution to A_{FB} which is of pure QED origin is not included because it is model independent and already respected in the experimental data. The sum of the QED corrections to Z exchange and γZ interference and the purely weak corrections turn out to be very small over the parameter range considered for realistic cuts ($\leqq 0.001$ in A_{FB}). Fig. 29 shows the predictions for $A_{FB}(|\cos\theta| \leqq 0.8)$ for $\sqrt{s} = 34.5$ GeV with an accolinearity cut of 10^0 and an energy cut of $0.5E_{\text{beam}}$

Table 1

Purely weak corrections to $A_{FB}(|\cos\theta| < 1)$ in percent for $\sqrt{s} = 43$ GeV ($M_H = 100$ GeV, $m_t = 30$ GeV).

s_W^2	M_Z(GeV)								
	89	90	91	92	93	94	95	96	97
0.15	−1.66	−1.62	−1.57	−1.53	−1.49	−1.45	−1.42	−1.38	−1.35
0.16	−1.57	−1.53	−1.48	−1.44	−1.41	−1.37	−1.34	−1.30	−1.27
0.17	−1.49	−1.45	−1.41	−1.37	−1.33	−1.30	−1.27	−1.23	−1.20
0.18	−1.41	−1.37	−1.34	−1.30	−1.27	−1.23	−1.20	−1.17	−1.14
0.19	−1.35	−1.31	−1.28	−1.24	−1.21	−1.18	−1.15	−1.12	−1.09
0.20	−1.29	−1.26	−1.22	−1.19	−1.16	−1.13	−1.10	−1.07	−1.05
0.21	−1.24	−1.21	−1.17	−1.14	−1.11	−1.08	−1.06	−1.03	−1.00
0.22	−1.19	−1.16	−1.13	−1.10	−1.07	−1.04	−1.02	−0.99	−0.97
0.23	−1.15	−1.12	−1.09	−1.06	−1.03	−1.01	−0.98	−0.96	−0.93
0.24	−1.12	−1.08	−1.06	−1.03	−1.00	−0.97	−0.95	−0.93	−0.90
0.25	−1.08	−1.05	−1.02	−1.00	−0.97	−0.94	−0.92	−0.90	−0.88
0.26	−1.05	−1.02	−0.99	−0.97	−0.94	−0.92	−0.89	−0.87	−0.85
0.27	−1.02	−0.99	−0.97	−0.94	−0.92	−0.89	−0.87	−0.85	−0.83
0.28	−1.00	−0.97	−0.94	−0.92	−0.89	−0.87	−0.85	−0.83	−0.81
0.29	−0.97	−0.95	−0.92	−0.90	−0.87	−0.85	−0.83	−0.81	−0.79
0.30	−0.95	−0.92	−0.90	−0.88	−0.85	−0.83	−0.81	−0.79	−0.77

for the bremsstrahlung part. These results are scaled up to $|\cos\theta| \leqq 1$ according to the lowest order formula (7.15) in order to be directly comparable with the PETRA results [4] that are shown in the figure, too.

One can see from fig. 29 that the experimental result for A_{FB} at 34.5 GeV favours values of s_W^2 which are slightly smaller than those following from the M_W/M_Z ratio.

d) Tests of the standard model at the 1-loop level:

The most important parameters of the electroweak standard model are the masses of the intermediate bosons. Therefore, we have expressed every observable quantity with help of M_W and M_Z. Consequently each measured value R_{exp} of a quantity R

$$R^{\text{exp}} = R^{\text{theor}}(M_W, M_Z) \tag{7.29}$$

gives a relation between M_W and M_Z. This depends if one includes radiative corrections slightly on the other parameters of the model i.e. the fermion masses and the Higgs mass. In order to perform a test of the standard model using purely leptonic reactions we

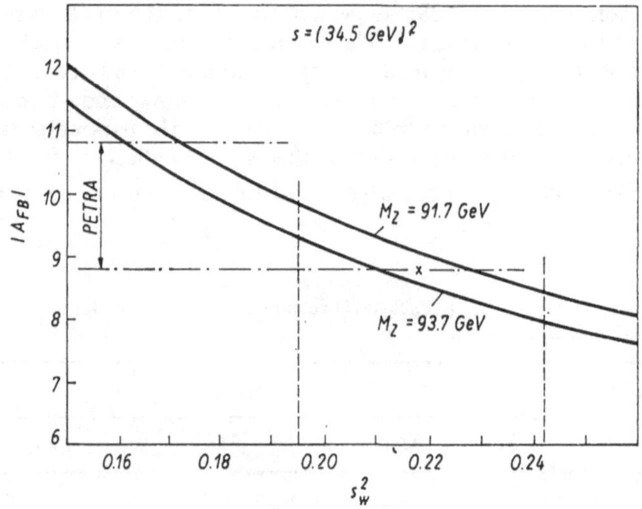

Fig. 29. A_{FB} (0.8) scaled up to $|\cos \theta| \leq 1$ according to eq. (7.15) at $\sqrt{s} = 34.5$ GeV as function of $s_W{}^2$ including complete electroweak radiative corrections. Shown are the curves resulting from the upper (93.7) and lower (91.7) bounds on M_Z from the $P\bar{P}$ collider experiment together with the upper and lower bounds on $s_W{}^2$ (dashed lines) from the same experiment. The cross marks the point corresponding to the mean values $M_Z = 92.7$ GeV, $s_W{}^2 = 0.218$. The dashed-dotted lines mark the PETRA results

present the experimental results in the (M_W, M_Z) plane. This is done in fig. 30 including 1-loop corrections (for our previously specified standard set of parameters (6.2)) for:

- μ decay, which gives a curve in the (M_W, M_Z) plane;
- $\overset{(-)}{\nu}_\mu e$ scattering, yielding relatively weak bounds on M_W, M_Z due to the present experimental errors [30];
- the lepton pair forward-backward asymmetry at $\sqrt{s} = 34.5$ GeV;
- the direct measurement of M_W and M_Z in the $P\bar{P}$ collider [1].

This picture represents a comprehensive test of the electroweak standard theory at the 1-loop level in the leptonic sector.

Clearly the low energy data (from Γ_μ, R_ν, A_{FB}) and the high energy data (M_W, M_Z) are compatible with each other. The agreement would be worse if radiative corrections were not taken into account. But in order to become really sensitive to these corrections improvements in the experimental determination of M_W, M_Z and R_ν are necessary.

8. Conclusion and Outlook

In this paper we have worked out a renormalization scheme for the standard electroweak model characterized by the following properties: use of the electric charge e and particle masses as physical parameters; minimal number of field renormalization constants respecting the $SU(2) \times U(1)$ symmetry; the simple pole structure of the 't Hooft-Feynman gauge is maintained after renormalization in a way consistent with the Slavnov-Taylor identities. We have calculated all the physical and unphysical self and mixing energies together with the complete set of renormalization constants, and the fermion gauge boson vertices. We have presented also the renormalized results for the self energies, vertices and box diagrams. These are the building blocks needed for the calculation

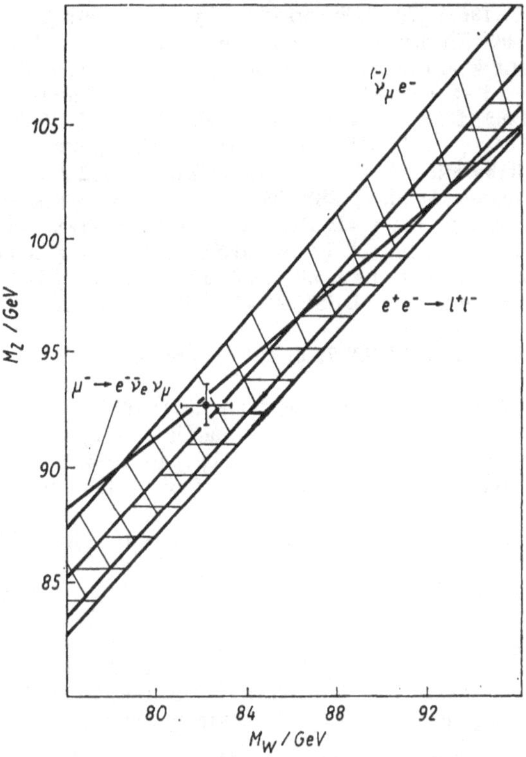

Fig. 30. Comparison of the results for the boson masses in the (M_W, M_Z) plane. Shown are: the curve resulting from μ decay (——), the 68% CL band determined from ν/e scattering (///////) and that from the forward-backward asymmetry in $e^+e^- \to \mu^+\mu^-$ (≡≡≡). The blob with the error bars represents the combined UA 1 and UA 2 results

of electroweak radiative corrections to e^+e^- annihilation, deep inelastic lepton scattering and $P\overline{P}$ annihilation at high energies.

Finally we have performed a comparison between experiment and the results of the standard model including radiative corrections for μ-decay, $\overset{(-)}{\nu}_\mu$e-scattering and e^+e^--annihilation into lepton pairs. We find good agreement between theory and experiment if radiative corrections are included. But the accuracy of the experiment has not yet reached a level where radiative corrections can be conclusively tested. We expect this to be the case in the near future, when results of the next generation of the e^+e^--machines and of the high precision experiments on lepton scattering are available. These then may also help to gather some information on the Higgs sector and finally will allow to discriminate between the minimal electroweak model and more extended versions.

References

[1] UA 1 Collaboration, G. Arnison et al., Phys. Lett. **126 B** (1983) 398; UA 2 Collaboration, P. Bagnaia et al., Phys. Lett. **129 B** (1983) 130; UA 1 Collaboration, G. Arnison et al., Phys. Lett. **129 B** (1983) 273; E. Radermacher, CERN-EP/84-41; L. DiLella, HEP 85 Conference, Bari 1985.

[2] S. L. GLASHOW, Nucl. Phys. **22** (1961) 579; S. WEINBERG, Phys. Rev. Lett. **19** (1967) 1264; A. SALAM, in Elementary Particle Theory, ed. N. Svartholm (Almqvist and Wiksell, Stockholm 1968), p. 367; S. L. GLASHOW, J. ILLIOPOULOS, L. MAIANI, Phys. Rev. **D 2** (1970) 1285.
[3] R. H. HEISTERBERG et al., Phys. Rev. Lett. **44** (1979) 635; CHARM Collaboration, F. BERGSMA et al., Phys. Lett. **117 B** (1982) 272; W. KRENZ, Aachen PITHA 82/26 (1982); L. A. AHRENS et al., Phys. Rev. Lett. **51** (1983) 1514; CDHS Collaboration, CH. GEWENIGER, Brighton Conference 1983; CHARM Collaboration, M. JONKER et al., CERN-EP/82-207 (1982).
[4] PLUTO Collaboration, CH. BERGER et al., DESY 83-084; P. GROSSE-WIESMANN, DESY 83-087; TASSO Collaboration, M. ALTHOFF et al., DESY 83-089; G. HERTEN, Brighton Conference 1983; E. LOHRMANN, DESY 83-102; A. BÖHM, DESY 83-103; B. NAROSKA, DESY 83-111; E. FERNANDEZ et al., Phys. Rev. Lett. **50** (1983) 1238; SLAC-PUB-3133 (1983); H. V. MARTYN, Vanderbilt Conference 1984, DESY 84-48; I. P. DUERDOTH, HEP '85 Conference, Bari 1985.
[5] Proceedings of the LEP Summer Study, CERN 79-01 (1979); Proceedings of the SLC Workshop, SLAC-Report-247 (1982).
[6] G. 'T HOOFT, Nucl. Phys. **B 33** (1971) 173, Nucl. Phys. **B 35** (1971) 167.
[7] Proceedings of the Topical Conference on Radiative Corrections in $SU(2)_L \times U(1)$, Miramare-Trieste, 6—8 June 1983, ed. B. W. Lynn and J. Wheater, Singapore 1984.
[8] D. YU. BARDIN, P. CH. CHRISTOVA, O. M. FEDERENKO, Nucl. Phys. **B 175** (1980) 435; **B 197** (1982) 1.
[9] G. PASSARINO, M. VELTMAN, Nucl. Phys. **B 160** (1979) 151; M. CONSOLI, Nucl. Phys. **B 160** (1979) 208.
[10] A. SIRLIN, Phys. Rev. **D 22** (1980) 971; W. J. MARCIANO, A. SIRLIN, Phys. Rev. **D 22** (1980) 2695; A. SIRLIN, W. J. MARCIANO, Nucl. Phys. **B 189** (1981) 442.
[11] J. FLEISCHER, F. JEGERLEHNER, Phys. Rev. **D 23** (1981) 2001.
[12] D. A. ROSS, J. C. TAYLOR, Nucl. Phys. **B 51** (1973) 25; S. SAKAKIBARA, Phys. Rev. **D 24** (1981) 1149; E. A. PASCHOS, M. WIRBEL, Nucl. Phys. **B 194** (1982) 189; M. WIRBEL, Z. Phys. **C 14** (1982) 293; I. LIEDE, E. A. PASCHOS, M. ROOS, S. SAKAKIBARA, Preprint HU-TFT-83-45/DO-TH-83-21 (1983).
[13] F. ANTONELLI, G. CORBO, M. CONSOLI, O. PELLEGRINO, Nucl. Phys. **B 183** (1981) 475; M. CONSOLI, S. LO PRESTI, L. MAIANI, Nucl. Phys. **B 223** (1983) 474.
[14] C. H. LLEWELLYN SMITH, J. F. WHEATER, Phys. Lett. **105 B** (1981) 486; J. F. WHEATER, C. H. LLEWELLYN SMITH, Nucl. Phys. **B 208** (1982) 27.
[15] K. I. AOKI, Z. HIOKI, R. KAWABE, M. KONUMA, T. MUTA, Progr. Theor. Phys. **64** (1980) 707; 65 (1981) 1001; Suppl. Progr. Theor. Phys. **73** (1982) 1.
[16] W. WETZEL, Nucl. Phys. **B 227** (1983) 1; R. W. BROWN, R. DECKER, E. A. PASCHOS, Phys. Rev. Lett. **52** (1984) 1192.
[17] J. COLE, in [7], and Nucl. Phys. B (to be published).
[18] M. GRECO, G. PANCHERI-SRIVASTAVA, Y. SRIVASTAVA, Nucl. Phys. **B 171** (1980) 118; F. A. BERENDS, R. KLEISS, S. JADACH, Nucl. Phys. **B 202** (1982) 63; M. BÖHM, W. HOLLIK, Nucl. Phys. **B 204** (1982) 45
[19] M. BÖHM, W. HOLLIK, Z. Phys. **C 23** (1984) 31.
[20] M. BÖHM, W. HOLLIK, Phys. Lett. **139 B** (1984) 213.
[21] L. D. FADDEEV, V. N. POPOV, Phys. Lett. **25 B** (1967) 29.
[22] C. BECCHI, A. ROUET, R. STORA, Phys. Lett. **52 B** (1974) 344; Comm. Math. Phys. **42** (1975) 127.
[23] A. A. SLAVNOV, Theor. Math. Phys. **10** (1972) 99; J. C. TAYLOR, Nucl. Phys. **B 33** (1971) 436.
[24] J. C. WARD, Phys. Rev. **78** (1950) 1824.
[25] G. 'T HOOFT, M. VELTMAN, Nucl. Phys. **B 44** (1972) 189.
[26] Particle Data Group, Phys. Lett. **111 B** (1982).
[27] J. GASSER, H. LEUTWYLER, Ann. Phys. (N.Y.) **136** (1981) 62; Phys. Rep. **87 C** (1982) 77.
[28] F. A. BERENDS, G. J. KOMEN, Phys. Lett. **63 B** (1976) 432. F. YNDURAIN, Nucl. Phys. **B 136**, (1978) 533.
[29] M. BÖHM, W. HOLLIK, H. SPIESBERGER, Z. Phys. **C 27** (1985) 523.
[30] CHARM Collaboration, F. BERGSMA et al. Contributed paper, Neutrino 84, Nordkirchen near Dortmund (1984); M. Murtagh, this conference. Paper submitted to the HEP '85 Conference, Bari 1985.

[31] M. Green, M. Veltman, Nucl. Phys. **B 169** (1980) 137; E: **B 175** (1980) 547; M. Veltman, Phys. Lett. **91 B** (1980) 95.
[32] M. Böhm, A. Denner, W. Hollik, R. Sommer, Phys. Lett. **144 B** (1984) 414.
[33] A. Denner, Diploma Thesis, Würzburg 1984.

Appendix A: Feynman rules and counter terms

We present the Feynman rules of the standard model using e, M_W, M_Z, M_H, $m_{i\sigma}$ as parameters and

$$c = M_W/M_Z, \quad s = (1 - M_W{}^2/M_Z{}^2)^{1/2}, \quad v_{i\sigma} = (I^3_{i\sigma} - 2s^2Q_{i\sigma})/2sc, \quad a_{i\sigma} = I^3_{i\sigma}/2sc$$

as abbreviations for writing out the couplings (again we suppress the index W at c_W, s_W). We combine the renormalization constants $\delta Z_i{}^W$, $\delta Z_i{}^B$, $\delta\xi_i{}^3$, $\delta\xi_i{}^B$, $\delta\tilde{Z}^W$, $\delta\tilde{Z}^B$ to those for the photon, Z boson and mixing terms:

$$\begin{aligned}
&\delta Z_i{}^\gamma = s^2\delta Z_i{}^W + c^2\delta Z_i{}^B, \qquad \delta\xi_i{}^\gamma = s^2\delta\xi_i{}^3 + c^2\delta\xi_i{}^B,\\
&\delta Z_i{}^Z = c^2\delta Z_i{}^W + s^2\delta Z_i{}^B, \qquad \delta\xi_i{}^Z = c^2\delta\xi_i{}^3 + s^2\delta\xi_i{}^B,\\
&\delta Z_i{}^{\gamma Z} = cs(\delta Z_i{}^W - \delta Z_i{}^B) = \frac{cs}{c^2 - s^2}(\delta Z_i{}^Z - \delta Z_i{}^\gamma),\\
&\delta\xi_i{}^{\gamma Z} = cs(\delta\xi_i{}^3 - \delta\xi_i{}^B) = \frac{cs}{c^2 - s^2}(\delta\xi_i{}^Z - \delta\xi_i{}^\gamma),
\end{aligned} \tag{A.1}$$

similar for $\delta\tilde{Z}$.

The renormalization constants $\delta M_W{}^2$, $\delta M_Z{}^2$, $\delta M_H{}^2$, $\delta m_{i\sigma}$, δt are defined in eq.s (4.1′) and (4.4′). This gives the following list of Feynman rules and counter terms (momenta and charges are understood as incoming):

$$= i\frac{2s}{e}M_WM_H{}^2\frac{\delta t}{t},$$

$$= \frac{i}{k^2 - M^2}\{-g_{\mu\nu}; 1; 1; \not k + m_{i\sigma}\},$$

$$= -ig_{\mu\nu}[(k^2 - M_a{}^2)\,\delta Z_2{}^a - \delta M_a{}^2] - ik_\mu k_\nu\delta\xi_1{}^a; \quad a = \pm, Z, \gamma,$$

$$= -ig_{\mu\nu}[-k^2\delta Z_2{}^{\gamma Z} + M_Z{}^2(\delta Z_1{}^{\gamma Z} - \delta Z_2{}^{\gamma Z})] + ik_\mu k_\nu\delta\xi_1{}^{\gamma Z},$$

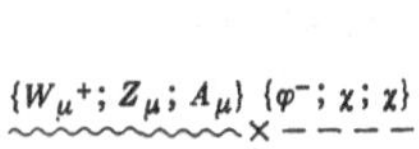

$$= k_\mu\cdot\frac{1}{2}M_{W,Z}\{i(\delta\xi_1{}^W - \delta\xi_2{}^W); -\delta\xi_1{}^Z + \delta\xi_2{}^Z; +\delta\xi_1{}^{\gamma Z} - \delta\xi_2{}^{\gamma Z}\},$$

$$= i[(k^2 - M_H{}^2)\,\delta Z^\varphi - \delta M_H{}^2],$$

$$= i\left[(k^2 - M^2_{W,Z})\,\delta Z^\varphi - \delta M^2_{W,Z} - M^2_{W,Z}\delta\xi_2{}^{W,Z} + M_H{}^2\frac{\delta t}{t}\right],$$

$$= i\left[k^2\left(\delta\tilde{Z}^a - \frac{1}{2}\,\delta\xi_1{}^a\right) - M_a{}^2\left(\delta\tilde{Z}^a + \frac{\delta M_a{}^2}{M_a{}^2} + \frac{1}{2}\,\delta\xi_2{}^a\right)\right];$$

$$a = \pm, Z, \gamma,$$

$$= -i\left[k^2\left(\delta\tilde{Z}^{\gamma Z} - \frac{1}{2}\,\delta\xi_1{}^{\gamma Z}\right) + M_Z{}^2\left(\delta Z_1{}^{\gamma Z} - \frac{3}{2}\,\delta Z_2{}^{\gamma Z} + \delta\tilde{Z}^{\gamma Z}\right)\right],$$

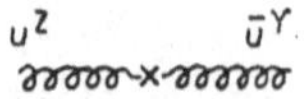

$$= -i\left[k^2\left(\delta\tilde{Z}^{\gamma Z} - \frac{1}{2}\,\delta\xi_1{}^{\gamma Z}\right) + M_Z{}^2\left(\delta Z_1{}^{\gamma Z} - \frac{3}{2}\,\delta Z_2{}^{\gamma Z} + \frac{1}{2}\,\delta\xi_2{}^{\gamma Z}\right)\right],$$

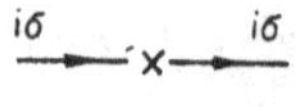

$$= i\left[\not{k}\left(\delta Z_L{}^i\,\frac{1-\gamma_5}{2} + \delta Z_R{}^{i\sigma}\,\frac{1+\gamma_5}{2}\right)\right.$$
$$\left. - m_{i\sigma}\left(\frac{1}{2}\,\delta Z_L{}^i + \frac{1}{2}\,\delta Z_R{}^{i\sigma}\right) - \delta m_{i\sigma}\right],$$

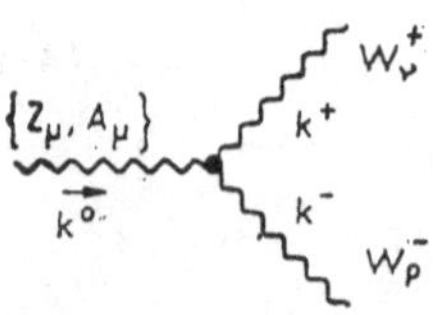

$$= -ie\left\{-\frac{c}{s};\ 1\right\}\left[g_{\mu\nu}(k^+ - k^0)_\varrho + g_{\nu\varrho}(k^- - k^+)_\mu\right.$$
$$\left. + g_{\varrho\mu}(k^0 - k^-)_\nu\right](1 + \delta Z_1{}^W),$$

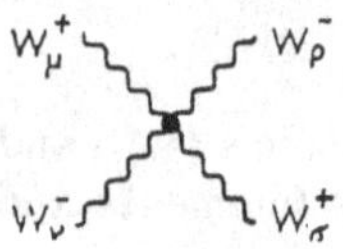

$$= ie^2\,\frac{1}{s^2}\left[2g_{\mu\sigma}g_{\nu\varrho} - g_{\mu\nu}g_{\varrho\sigma} - g_{\mu\varrho}g_{\nu\sigma}\right]\cdot(1 + 2\delta Z_1{}^W - \delta Z_2{}^W),$$

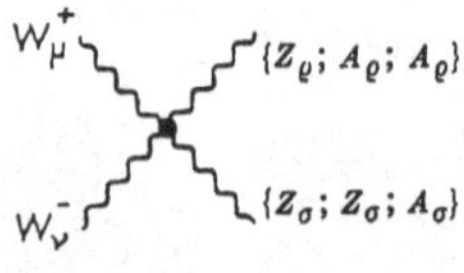

$$= -ie^2\left\{\frac{c^2}{s^2};\ -\frac{c}{s};\ 1\right\}\left[2g_{\mu\nu}g_{\varrho\sigma} - g_{\mu\varrho}g_{\nu\sigma} - g_{\mu\sigma}g_{\nu\varrho}\right]$$
$$\times\,(1 + 2\delta Z_1{}^W - \delta Z_2{}^W),$$

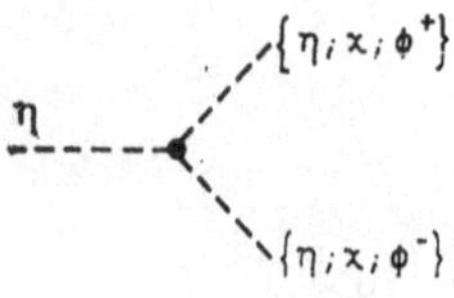

$$= -i\,\frac{e}{2s}\,\frac{M_H{}^2}{M_W}\,\{3;\ 1;\ 1\}\left(1 - \frac{\delta v}{v} + \delta Z^\lambda\right),$$

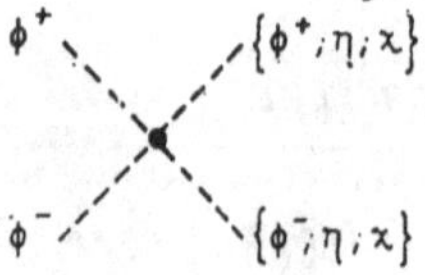

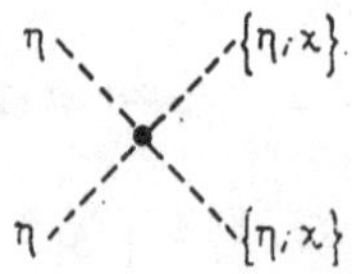

;
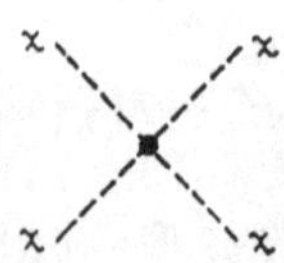

$$= -i\,\frac{e^2}{4s^2}\,\frac{M_H{}^2}{M_W{}^2}\,\{2;\ 1;\ 1;\ 3;\ 1;\ 3\}$$
$$\times\,(1 + \delta Z^\lambda),$$

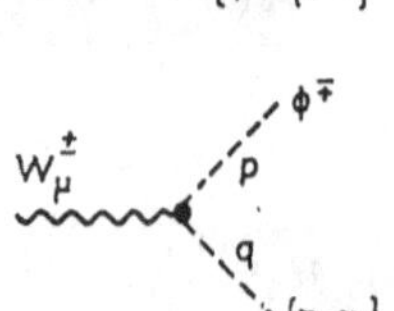

$$= \frac{e}{2s}\,(p - q)_\mu\,\{\mp i;\ 1\}\,(1 + \delta Z^\varphi + \delta Z_1{}^W - \delta Z_2{}^W),$$

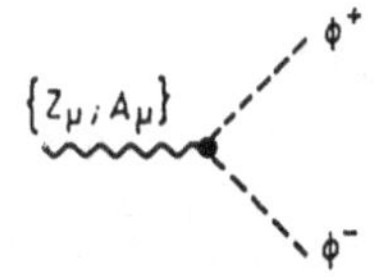

$$= -ie(p-q)_\mu \left\{\frac{s^2-c^2}{2cs};\, 1\right\} \Bigg(1 + \delta Z^\varphi + \left\{\delta Z_1{}^Z - \delta Z_2{}^Z + \frac{2cs}{c^2-s^2}(\delta Z_1{}^{\gamma Z} - \delta Z_2{}^{\gamma Z});\right.$$
$$\left.\delta Z_1{}^\gamma - \delta Z_2{}^\gamma + \frac{c^2-s^2}{2cs}(\delta Z_1{}^{\gamma Z} - \delta Z_2{}^{\gamma Z})\right\}\Bigg),$$

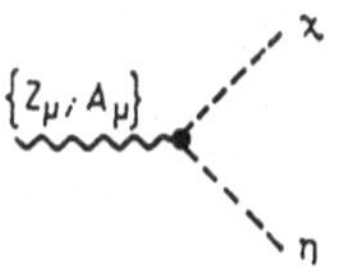

$$= \frac{e}{2cs}(p-q)_\mu \{1 + \delta Z^\varphi + \delta Z_1{}^Z - \delta Z_2{}^Z;\, \delta Z_2{}^{\gamma Z} - \delta Z_1{}^{\gamma Z}\},$$

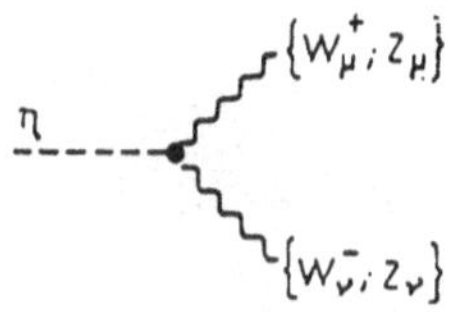

$$= i\frac{e}{s} M_W g_{\mu\nu} \left\{1;\, \frac{1}{c^2}\right\} \left(1 + \frac{1}{2}\delta Z^\varphi + \frac{\delta M_{W;Z}}{M_{W;Z}} + \delta Z_1{}^{W;Z} - \frac{1}{2}\delta Z_2{}^{W;Z}\right),$$

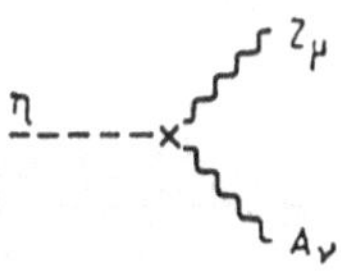

$$= -i\frac{e}{c^2 s} M_W g_{\mu\nu}(\delta Z_1{}^{\gamma Z} - \delta Z_2{}^{\gamma Z}),$$

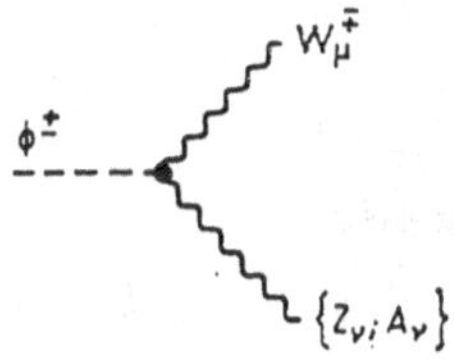

$$= -ieM_W g_{\mu\nu} \left\{\frac{s}{c};\, 1\right\} \left(1 - \frac{1}{2}\delta Z^\varphi + \frac{\delta M_W}{M_W} + \frac{1}{2}\delta Z_2{}^W + \delta Z_1{}^B - \delta Z_2{}^B\right),$$

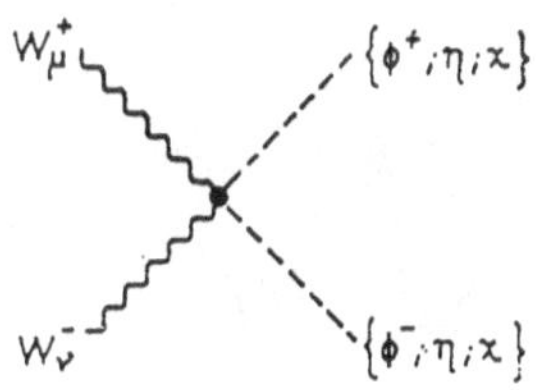

$$= i\frac{e^2}{2s^2} g_{\mu\nu}(1 + \delta Z^\varphi + 2\delta Z_1{}^W - 2\delta Z_2{}^W),$$

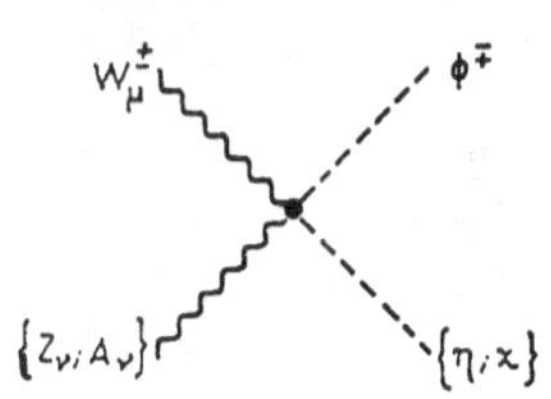

$$= -i\frac{e^2}{2} g_{\mu\nu} \left\{\frac{1}{c};\, \frac{1}{s}\right\} \{1;\, \pm i\}\, (1 + \delta Z^\varphi + \delta Z_1{}^W - \delta Z_2{}^W + \delta Z_1{}^B - \delta Z_2{}^B),$$

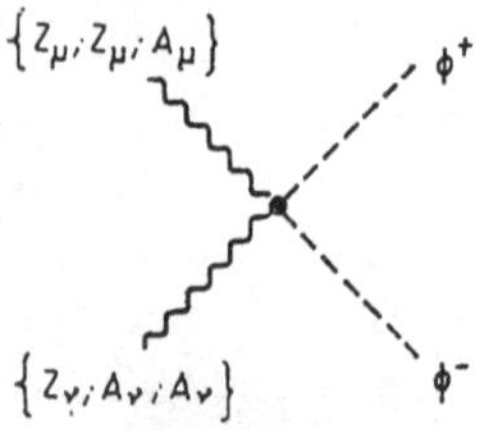

$$= ie^2 g_{\mu\nu} \begin{cases} \dfrac{(c^2-s^2)^2}{2c^2s^2}\Big(1 + \delta Z^\varphi + 2\delta Z_1{}^Z - 2\delta Z_2{}^Z \\ \qquad - \dfrac{4cs}{c^2-s^2}(\delta Z_1{}^{\gamma Z} - \delta Z_2{}^{\gamma Z})\Big), \\ \dfrac{s^2-c^2}{cs}\Big(1 + \delta Z^\varphi + \delta Z_1{}^\gamma - \delta Z_2{}^\gamma + \delta Z_1{}^Z - \delta Z_2{}^Z \\ \qquad + \dfrac{1}{2cs(c^2-s^2)}(\delta Z_1{}^{\gamma Z} - \delta Z_2{}^{\gamma Z})\Big) \\ 2\Big(1 + \delta Z^\varphi + 2\delta Z_1{}^\gamma - 2\delta Z_2{}^\gamma \\ \qquad + \dfrac{c^2-s^2}{cs}(\delta Z_1{}^{\gamma Z} - \delta Z_2{}^{\gamma Z})\Big), \end{cases}$$

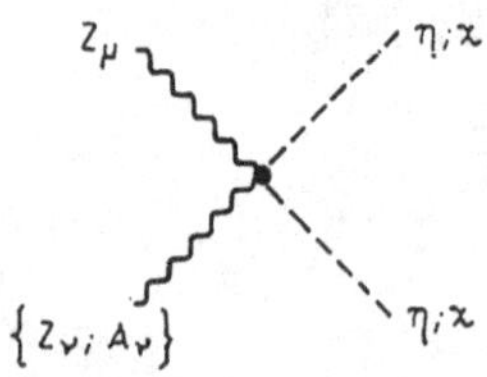

$$= i\frac{e^2}{2c^2s^2} g_{\mu\nu}\{1 + 2\delta Z_1{}^Z - 2\delta Z_2{}^Z + \delta Z^\varphi; \delta Z_2{}^{\gamma Z} - \delta Z_1{}^{\gamma Z}\},$$

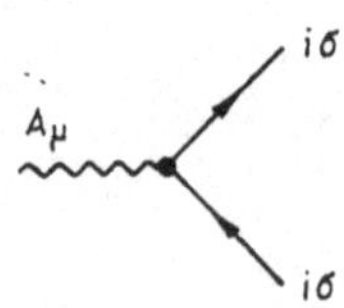

$$= -ieQ^{i\sigma}\gamma_\mu\left(1 + \delta Z_1{}^\gamma - \delta Z_2{}^\gamma + \delta Z_L{}^i \frac{1-\gamma_5}{2} + \delta Z_R{}^{i\sigma}\frac{1+\gamma_5}{2}\right)$$
$$- ie\gamma_\mu(v_{i\sigma} - \gamma_5 a_{i\sigma})(\delta Z_1{}^{\gamma Z} - \delta Z_2{}^{\gamma Z}),$$

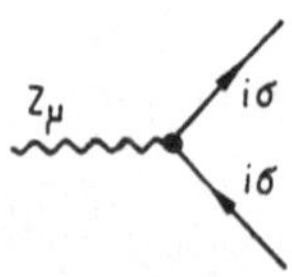

$$= ie\gamma_\mu(v_{i\sigma} - \gamma_5 a_{i\sigma})(1 + \delta Z_1{}^Z - \delta Z_2{}^Z)$$
$$+ i\frac{e}{cs}\gamma_\mu\Big[(I_3{}^\sigma - s^2Q^{i\sigma}\frac{1-\gamma_5}{2}\delta Z_L{}^i - s^2Q^{i\sigma}\frac{1+\gamma_5}{2}\delta Z_R{}^{i\sigma}$$
$$+ csQ^{i\sigma}(\delta Z_1{}^{\gamma Z} - \delta Z_2{}^{\gamma Z})\Big],$$

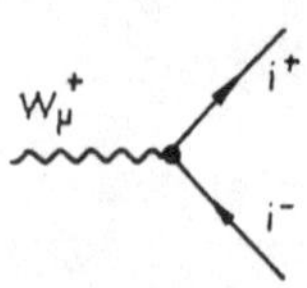

$$= i\frac{e}{\sqrt{2}\,s}\gamma_\mu\frac{1-\gamma_5}{2}(1 + \delta Z_1{}^W - \delta Z_2{}^W + \delta Z_L{}^i),$$

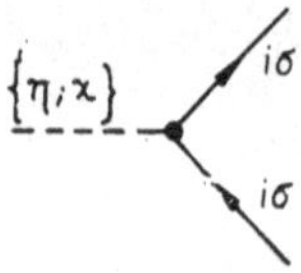

$$= -\frac{e}{2s}\frac{m_{i\sigma}}{M_W}\{i; 2I_3{}^\sigma\gamma_5\}\left(1 + \frac{\delta m_{i\sigma}}{m_{i\sigma}} + \frac{\delta v}{v}\right.$$
$$\left. + \frac{1}{2}\delta Z_L{}^i + \frac{1}{2}\delta Z_R{}^{i\sigma}\right),$$

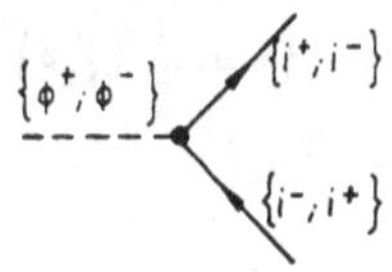

$$= i\,\frac{e}{\sqrt{2}\,s}\,\frac{1}{M_W}\left(\frac{1 \mp \gamma_5}{2}\,m_{i+}\left(1 + \frac{\delta m_{i+}}{m_{i+}} + \frac{\delta v}{v} + \frac{1}{2}\,\delta Z_R{}^{i+} + \frac{1}{2}\,\delta Z_L{}^{i}\right) - \frac{1 \pm \gamma_5}{2}\,m_{i-} \times \left(1 + \frac{\delta m_{i-}}{m_{i-}} + \frac{\delta v}{v} + \frac{1}{2}\,\delta Z_R{}^{i-} + \frac{1}{2}\,\delta Z_L{}^{i}\right)\right),$$

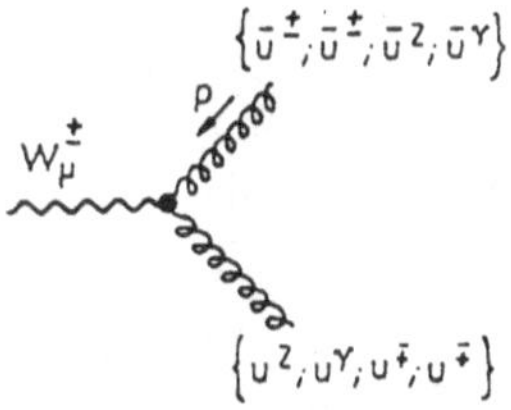

$$= \pm iep_\mu \left\{\frac{c}{s}; -1; -\frac{c}{s}; 1\right\}\left(1 + \delta\tilde{Z}^W + \delta Z_1{}^W - \delta Z_2{}^W - \frac{1}{2}\,\{\delta\xi_1{}^W; \delta\xi_1{}^W; \delta\xi_1{}^3; \delta\xi_1{}^3\}\right),$$

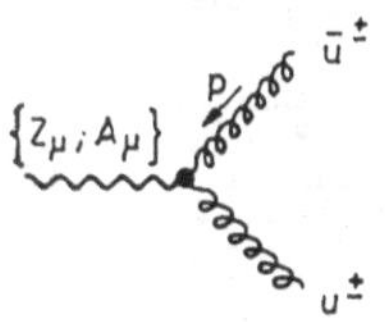

$$= \mp iep_\mu \left\{\frac{c}{s}; -1\right\}\left(1 + \delta\tilde{Z}^W - \frac{1}{2}\,\delta\xi_1{}^W + \delta Z_1{}^W - \delta Z_2{}^W\right),$$

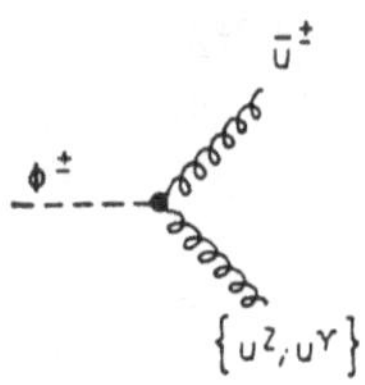

$$= ieM_W \left\{\frac{s^2 - c^2}{2cs}; 1\right\}\left(1 - \frac{\delta v}{v} + \delta Z^\varphi + \delta Z_1{}^W - \frac{3}{2}\,\delta Z_2{}^W + \left\{\frac{c^2}{c^2 - s^2}; \frac{1}{2}\right\}\left(\delta Z_1{}^W - \frac{3}{2}\,\delta Z_2{}^W + \delta\tilde{Z}^W + \delta\xi_2{}^W\right) + \left\{\frac{-s^2}{c^2 - s^2}; \frac{1}{2}\right\}\left(\delta Z_1{}^B - \frac{3}{2}\,\delta Z_2{}^B + \delta\tilde{Z}^B + \delta\xi_2{}^W\right)\right),$$

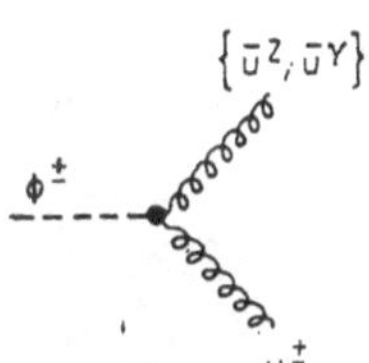

$$= i\,\frac{e}{2cs}\,M_W \left\{1 - \frac{\delta v}{v} + \delta Z^\varphi + \delta Z_1{}^Z - \frac{3}{2}\,\delta Z_2{}^Z + \delta Z_1{}^W - \frac{3}{2}\,\delta Z_2{}^W + \delta\tilde{Z}^W + \frac{1}{2}\,\delta\xi_2{}^Z; -\delta Z_1{}^{\gamma Z} + \frac{3}{2}\,\delta Z_2{}^{\gamma Z} - \frac{1}{2}\,\delta\xi_2{}^{\gamma Z}\right\},$$

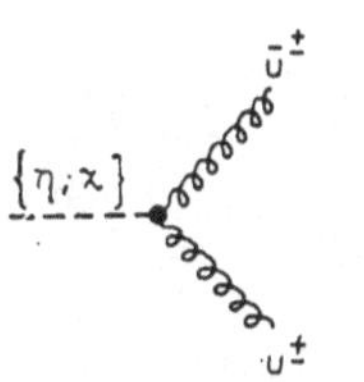

$$= -i\,\frac{e}{2s}\,M_W\{1; \pm i\}\left(1 - \frac{\delta v}{v} + \delta Z^\varphi + 2\delta Z_1{}^W - 3\delta Z_2{}^W + \delta\tilde{Z}^W + \frac{1}{2}\,\delta\xi_2{}^Z\right),$$

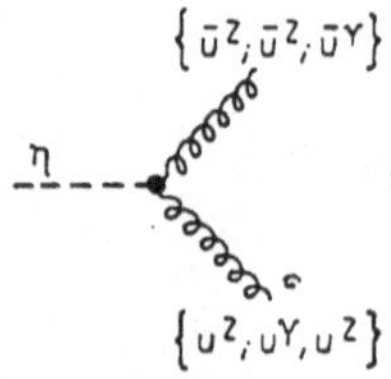

$$= -i \frac{e}{2c^2 s} M_W \cdot \begin{cases} \left(1 - \frac{\delta v}{v} + \delta Z^\varphi + 2\delta Z_1{}^Z - 3\delta Z_2{}^Z + \delta \tilde{Z}^Z + \frac{1}{2}\,\delta\xi_2{}^Z\right), \\ \left(-\delta Z_1{}^{\gamma Z} + \frac{3}{2}\,\delta Z_2{}^{\gamma Z} - \delta\tilde{Z}^{\gamma Z}\right), \\ \left(-\delta Z_1{}^{\gamma Z} + \frac{3}{2}\,\delta Z_2{}^{\gamma Z} - \frac{1}{2}\,\delta\xi_2{}^{\gamma Z}\right). \end{cases}$$

Appendix B:

1. The finite part of the *scalar self energy* $F(k^2; M_1, M_2)$:

$$F(k^2; M_1, M_2) = 1 + \left(\frac{M_1{}^2 - M_2{}^2}{k^2} - \frac{M_1{}^2 + M_2{}^2}{M_1{}^2 - M_2{}^2}\right) \ln \frac{M_2}{M_1}$$

$$+ \begin{cases} + \frac{1}{k^2}\left[\left((M_1 + M_2)^2 - k^2\right)\left((M_1 - M_2)^2 - k^2\right)\right]^{1/2} \\ \times \ln \frac{\sqrt{(M_1 + M_2)^2 - k^2} + \sqrt{(M_1 - M_2)^2 - k^2}}{\sqrt{(M_1 + M_2)^2 - k^2} - \sqrt{(M_1 - M_2)^2 - k^2}}, \\ k^2 < (M_1 - M_2)^2 \\ - \frac{2}{k^2}\left[(M_1 + M_2)^2 - k^2\right]^{1/2}\left[k^2 - (M_1 - M_2)^2\right]^{1/2} \\ \times \arctan \frac{\sqrt{k^2 - (M_1 - M_2)^2}}{\sqrt{(M_1 + M_2)^2 - k^2}}, \\ (M_1 - M_2)^2 < k^2 < (M_1 + M_2)^2 \\ - \frac{1}{k^2}\left[k^2 - (M_1 + M_2)^2\right]^{1/2}\left[k^2 - (M_1 - M_2)^2\right]^{1/2} \\ \times \left\{\ln \frac{\sqrt{k^2 - (M_1 + M_2)^2} + \sqrt{k^2 - (M_1 - M_2)^2}}{\sqrt{k^2 - (M_1 - M_2)^2} - \sqrt{k^2 - (M_1 + M_2)^2}} - i\pi\right\}, \\ k^2 > (M_1 + M_2)^2. \end{cases} \tag{B.1}$$

2. *The function* $B_1(k^2; M_1, M_2)$ is defined by:

$$\frac{i}{16\pi^2}\, k_\mu\, B_1(k^2; M_1, M_2) = \mu^{4-D} \int \frac{d^D q}{(2\pi)^D} \frac{q_\mu}{(q^2 - M_1{}^2 + i\varepsilon)\left((q + k^2) - M_2{}^2 + i\varepsilon\right.}$$

and related to B_0:

$$2k^2 B_1(k^2; M_1, M_2) = A(M_2) - A(M_1) + (M_2{}^2 - M_1{}^2 - k^2)\, B_0(k^2; M_1, M_2). \tag{B.2}$$

This gives for equal masses:

$$B_1(k^2; M, M) = -\frac{1}{2}\, B_0(k^2, M, M). \tag{B.2'}$$

3. *The vertex functions* $\Lambda_{1,\dots,4}$:

The contributions of the photonic diagrams of fig. 7a to the vertex function Γ_μ (of sect. 5.2f) for $|k^2| \gg m^2$, $\not{p} = \not{q} = m$ have already been calculated [18, 19]. After splitting off the UV divergent parts and the coupling constants remains the function $\Lambda_1(k^2, m)$:

$$\Lambda_1(0, m) = 0,$$
$$\Lambda_1(k^2, m) = -2 \ln \frac{|k^2|}{\lambda^2} \left(\ln \frac{|k^2|}{m^2} - 1\right) + \ln \frac{|k^2|}{m^2} + \ln^2 \frac{|k^2|}{m^2}$$
$$+ \begin{cases} 4\left(\dfrac{\pi^2}{3} - 1\right) + 2\pi i \left(\ln \dfrac{k^2}{\lambda^2} - \dfrac{3}{2}\right) & \text{for} \quad k^2 \gg m^2, \\ 4\left(\dfrac{\pi^2}{12} - 1\right) & \text{for} \quad -k^2 \gg m^2. \end{cases} \tag{B.3}$$

The diagrams of fig. 7b describing the exchange of the heavy bosons Z and W lead to the following integral:

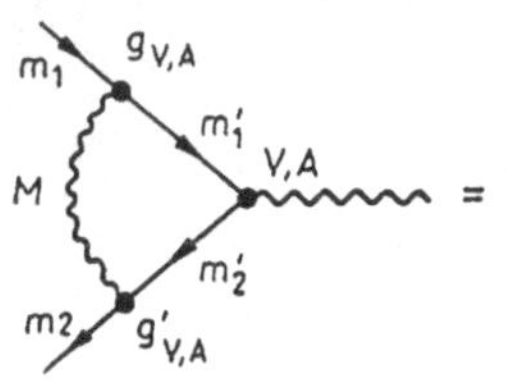

$$= \mu^{4-D} \int \frac{d^D l}{(2\pi)^D} \frac{\gamma_\nu(g_V' - g_A'\gamma_5)(-\not{q} - \not{l} + m_2')\gamma_\mu(V - A\gamma_5)(\not{p} - \not{l} + m_1')\gamma^\nu(g_V - g_A\gamma_5)}{(l^2 - M^2)\left((p - l)^2 + m_1'^2\right)\left((q + l)^2 - m_2'^2\right)}$$

$$= \frac{i}{16\pi^2} \gamma_\mu(\lambda_V - \lambda_A\gamma_5) \cdot \begin{cases} \left[\Delta_M - \dfrac{1}{2}\right] & \text{for} \quad k^2 = 0, \\ \left[\Delta_M - \dfrac{1}{2} + \Lambda_2(k^2, M)\right] & \text{for} \quad k^2 \gg m_{1,2}^2, m_{1,2}'^2 \end{cases}$$

with

$$\lambda_V = V(g_V g_V' + g_A g_A') + A(g_V g_A' + g_V' g_A),$$
$$\lambda_A = A(g_V g_V' + g_A g_A') + V(g_V g_A' + g_V' g_A),$$

and

$$\Lambda_2(k^2, M) = -\frac{5}{2} + \ln w + 2(1 + w)^2 \frac{d}{dw}\left(\frac{J(w)}{1 + w}\right), \quad w = \frac{M^2}{k^2}.$$

The parameter integral

$$J(w) = \int_0^1 dx \int_0^1 dy\, y \ln\left[w(1 - y) - y^2 x(1 - x) - i\varepsilon\right]$$

can be evaluated with help of the dilogarithm

$$\mathrm{Sp}(z) = -\int_0^1 dt \frac{\ln(1 - zt)}{t}$$

and yields for Λ_2 the expression:

$$\Lambda_2(k^2, M) = -\frac{7}{2} - 2w - (2w+3)\ln(-w) + 2(1+w)^2\left[\mathrm{Sp}\left(1+\frac{1}{w}\right) - \frac{\pi^2}{6}\right]$$
$$\text{for} \quad k^2 < 0,$$

$$\Lambda_2(k^2, M) = -\frac{7}{2} - 2w - (2w+3)\ln(w) + 2(1+w)^2\left[\ln(w)\ln\left(\frac{w+1}{w}\right) - \mathrm{Sp}\left(-\frac{1}{w}\right)\right] - i\pi\left[3 + 2w - 2(1+w)^2\ln\left(\frac{1+w}{w}\right)\right] \tag{B.4}$$
$$\text{for} \quad k^2 > 0.$$

In fig. 22 we show $\Lambda_2(k^2, M_Z)$, $\Lambda_2(k^2, M_W)$, $\Lambda_3(k^2, M_W)$, and $\Lambda_4(k^2; M_Z, M_W)$.

In a similar way we have obtained the invariant functions belonging to the diagram 7c containing the triple boson vertex:

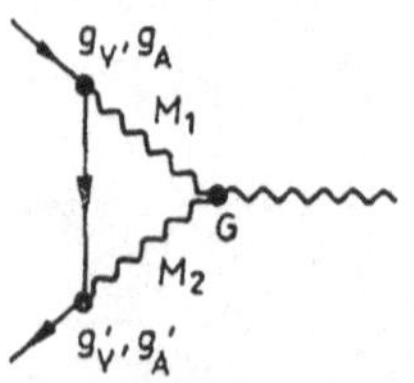

$$= \frac{i}{16\pi^2}\gamma_\mu(\lambda_1 - \lambda_2\gamma_5)\cdot\Lambda(k^2; M_1, M_2)$$

with

$$\lambda_1 = (g_V g_V' + g_A g_A')\,G, \qquad \lambda_2 = (g_V g_A' + g_A g_V')\,G$$

and

$$\Lambda(k^2; M_1, M_2) = \begin{cases} 3\left[\Delta_M - \frac{1}{6} + \Lambda_3(k^2, M)\right] & \text{for} \quad M_1 = M_2 = M, \\ 3\left[\frac{\Delta_{M_1} + \Delta_{M_2}}{2} + \frac{5}{6} - \frac{M_1^2 + M_2^2}{M_1^2 - M_2^2}\ln\frac{M_1}{M_2} + \Lambda_4(k^2; M_1, M_2)\right] & \text{for} \quad M_1 \neq M_2. \end{cases} \tag{B.5}$$

The remaining functions have the properties

$$\Lambda_3(0, M) = \Lambda_4(0, M_1, M_2) = 0,$$
$$\Lambda_4(k^2, M_1, M_2) = \Lambda_4(k^2, M_2, M_1).$$

They read for $|k^2| \gg m_f^2$ $\quad (w = M^2/k^2)$:

$$\Lambda_3(k^2, M) = \frac{5}{6} - \frac{2w}{3} + \frac{2w+1}{3}\sqrt{1-4w}\,\ln\frac{\sqrt{1-4w}+1}{\sqrt{1-4w}-1} + \frac{2}{3}w(w+2)\left(\ln\frac{\sqrt{1-4w}+1}{\sqrt{1-4w}-1}\right)^2 \qquad \text{for} \quad k^2 < 0,$$

$$\Lambda_3(k^2, M) = \frac{5}{6} - \frac{2w}{3} + \frac{2}{3}(2w+1)\sqrt{4w-1}\arctan\frac{1}{\sqrt{4w-1}}$$
$$-\frac{8}{3}w(w+2)\left(\arctan\frac{1}{\sqrt{4w-1}}\right)^2 \quad \text{for} \quad 0 < k^2 < 4M^2,$$
$$\tag{B.6}$$
$$\Lambda_3(k^2, M) = \frac{5}{6} - \frac{2w}{3} + \frac{2w+1}{3}\sqrt{1-4w}\ln\frac{1+\sqrt{1-4w}}{1-\sqrt{1-4w}}$$
$$+\frac{2}{3}w(w+2)\left[\ln^2\left(\frac{1+\sqrt{1-4w}}{1-\sqrt{1-4w}} - \pi^2\right)\right]$$
$$-i\pi\left[\frac{2w+1}{3}\sqrt{1-4w} + \frac{2}{3}w(w+2)\ln\frac{1+\sqrt{1-4w}}{1-\sqrt{1-4w}}\right]$$
$$\text{for} \quad k^2 > 4M^2$$

and with $w_1 = M_1^2/k^2$, $w_2 = M_2^2/k^2$:

$$\Lambda_4(k^2; M_1, M_2)$$
$$= \frac{1}{6} + \frac{w_1+w_2}{w_1-w_2}\ln\frac{M_1}{M_2} - \frac{w_1-w_2}{3}\ln\frac{M_1}{M_2} + \frac{w_1+w_2+1}{3}\left(\ln\frac{M_1}{M_2} - 1\right)$$
$$+\frac{w_1+w_2+1}{3}\left[x_1\ln\frac{x_1}{x_1-1} + x_2\ln\frac{-x_2}{1-x_2}\right]$$
$$-\frac{2}{3}(w_1+w_2+w_1w_2)\ln\frac{x_1}{x_1-1}\ln\frac{-x_2}{1-x_2}, \tag{B.7}$$

$$x_{1,2} = \begin{cases} \dfrac{1-w_1+w_2}{2} \pm \dfrac{1}{2}\sqrt{(1-w_1+w_2)^2 - 4w_2} \\ \text{for} \quad k^2 < (M_1-M_2)^2 \quad \text{and} \quad k^2 > (M_1+M_2)^2, \\ \dfrac{1-w_1+w_2}{2} \pm \dfrac{i}{2}\sqrt{4w_2 - (1-w_1+w_2)^2} \\ \text{for} \quad (M_1-M_2)^2 < k^2 < (M_1+M_2)^2. \end{cases}$$

The imaginary part of Λ_4 is obtained from (B.7).

$$\operatorname{Im}\Lambda_4(k^2; M_1, M_2) = -\pi\cdot\theta\big(k^2 - (M_1+M_2)^2\big)$$
$$\times\left\{\frac{w_1+w_2+1}{3}\sqrt{(1-w_1+w_2)^2-4w_2}\right.$$
$$\left.+\frac{2}{3}(w_1+w_2+w_1w_2)\left(\ln\frac{x_1}{1-x_1} + \ln\frac{x_2}{1-x_2}\right)\right\}. \tag{B.8}$$

In the case $M_2 = 0$, i.e. if one of the bosons is a photon, the mass m of the fermion coupled to the photon has to be respected. In that special case we have to use the following expression, valid for $|k^2| \gg m^2$ (with $w = M^2/k^2$):

$$\Lambda_4^m(k^2; M, 0) = \frac{1}{2} - \frac{w}{3} + \frac{1-w^2}{3} \ln \frac{w-1}{w} + \frac{2}{3} \ln \frac{M^2}{m^2}$$
$$+ \frac{w}{3} \left[\ln^2 \left(\frac{M^2 - k^2}{m^2} \right) - \left(\ln \frac{M^2}{m^2} \right)^2 + 2\,\mathrm{Sp} \left(\frac{1}{1-w} \right) \right]. \qquad \text{(B.9)}$$

For $k^2 \cong M^2$ replace $M^2 \to M^2 - iM\Gamma$.

Fig. 23 shows $\Lambda_4^m(k^2; M, 0)$ for $m = m_e$, $M = M_W$.

4. *The box diagrams:*

We write down the explicit expressions for the s-channel box diagrams. The matrix element can be written as:

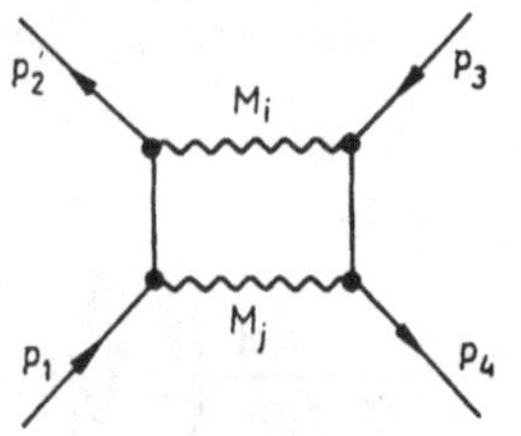

$$= \frac{\alpha}{2\pi} \frac{1}{s} \{ \bar{v}(p_2)\, \gamma_\mu (\lambda_1^V - \lambda_1^A \gamma_5)\, u(p_1) \cdot \bar{u}(p_4)\, \gamma^\mu (\lambda_2^V - \lambda_2^A \gamma_5)\, v(p_3) \cdot I^{ij}(s, t)$$
$$+ \bar{v}(p_2)\, \gamma_\mu \gamma_5 (\lambda_1^V - \lambda_1^A \gamma_5)\, u(p_1) \cdot \bar{u}(p_4)\, \gamma^\mu \gamma_5 (\lambda_2^V - \lambda_2^A \gamma_5)\, v(p_3) \cdot I_5^{ij}(s, t) \} \qquad \text{(B.10)}$$

with

$$\lambda_1^V = v_1 v_2 + a_1 a_2, \qquad \lambda_1^A = v_1 a_2 + v_2 a_1,$$
$$\lambda_2^V = v_3 v_4 + a_3 a_4, \qquad \lambda_2^A = v_3 a_4 + v_4 a_3,$$

where the v_i, a_i are the vector and axial vector couplings of the i'th fermion to the internal boson according to the labelling of fermion momenta in the above diagram.

$$s = (p_1 + p_2)^2, \qquad t = (p_2 - p_3)^2, \qquad u = (p_1 - p_3)^2$$

are the usual Mandelstam variables.

In order to obtain the expression for the crossed box one should substitute $t \leftrightarrow u$ and reverse the sign of I^{ij}:

$$I^{ij}(s, t) \to -I^{ij}(s, u) \quad \text{and} \quad I_5^{ij}(s, t) \to I_5^{ij}(s, u). \qquad \text{(B.11)}$$

The following results [33], given in a form which is convenient for numerical evaluation, are valid for s, $|t|$, $|u| \gg m_f^2$. Because of (B.11) we give for the neutral-current boxes only that part of I, which is antisymmetric under $t \leftrightarrow u$:

$$I_5^{\gamma\gamma}(s, t) = \frac{\alpha}{2\pi} \left\{ \frac{s}{2(s+t)} \ln \frac{t}{s + i\varepsilon} - \frac{s(s+2t)}{4(s+t)^2} \left[\ln^2 \left(\frac{t}{s + i\varepsilon} \right) + \pi^2 \right] \right\} \qquad \text{(B.12)}$$

$$I^{\gamma\gamma}(s, t) - I^{\gamma\gamma}(s, u) = \frac{\alpha}{2\pi} \left\{ -2 \ln \frac{t + i\varepsilon}{-s} \ln \left(\frac{-s}{\lambda^2} - i\varepsilon \right) \right\} + I_5^{\gamma\gamma}(s, t) - I_5^{\gamma\gamma}(s, u);$$

$$I_5^{\gamma Z}(s,t) = \frac{\alpha}{2\pi}\frac{s-M_Z^2}{2(s+t)}\left\{\ln\frac{t}{s-M_Z^2} - \frac{M_Z^2}{s}\ln\frac{M_Z^2}{M_Z^2-s} + \frac{s+2t+M_Z^2}{s+t}\right.$$
$$\left.\times\left[\mathrm{Sp}\frac{s}{M_Z^2} - \mathrm{Sp}\frac{-t}{M_Z^2} + \ln\frac{-t}{M_Z^2}\cdot\ln\frac{M_Z^2-s}{M_Z^2+t}\right]\right\}, \tag{B.13}$$

$$I^{\gamma Z}(s,t) - I^{\gamma Z}(s,u) = I_5^{\gamma Z}(s,t) - I_5^{\gamma Z}(s,u) + \frac{\alpha}{2\pi}\left(\mathrm{Sp}\frac{M_Z^2+t}{t}\right.$$
$$\left.+\frac{1}{2}\ln\frac{t^2}{s^2}\ln\frac{M_Z^2}{M_Z^2-s} - \ln\left(\frac{-t}{s}+i\varepsilon\right)\ln\frac{M_Z^2}{\lambda^2}\right).$$

$$I_5^{ZZ}(s,t) = \frac{\alpha}{2\pi}\frac{s}{s+t}\left\{\frac{2t+s+2M_Z^2}{2(s+t)}\left[\mathrm{Sp}\left(1+\frac{t}{M_Z^2}\right) - \frac{\pi^2}{6} - \left(\ln\frac{-y_1}{y_2}\right)^2\right]\right.$$
$$+\frac{1}{2}\ln\frac{-t}{M_Z^2} + \frac{y_2-y_1}{2}\ln\frac{-y_1}{y_2}$$
$$+\frac{s+2t-4tM_Z^2/s+2M_Z^4/t-2M_Z^4/s}{2(s+t)(x_2+x_1)}$$
$$\left.\times\left[\mathrm{Sp}\frac{x_1}{x_1-y_1} + \mathrm{Sp}\frac{x_1}{x_1-y_2} - \mathrm{Sp}\frac{x_2}{x_2-y_2} - \mathrm{Sp}\frac{x_2}{x_2-y_1}\right]\right\}, \tag{B.14}$$

$$I^{ZZ}(s,t) = I_5^{ZZ}(s,t) + \frac{\alpha}{2\pi}\left\{2\left(\ln\frac{-y_1}{y_2}\right)^2 + \frac{2}{x_1-x_2}\right.$$
$$\left.\times\left[\mathrm{Sp}\frac{x_1}{x_1-y_1} + \mathrm{Sp}\frac{x_1}{x_1-y_2} - \mathrm{Sp}\frac{x_2}{x_2-y_2} - \mathrm{Sp}\frac{x_2}{x_2-y_1}\right]\right\}$$

with

$$x_{1/2} = \frac{1}{2}\left(1 \pm \sqrt{1-4M_Z^2/s\,(1+M_Z^2/t)}\right)$$
$$y_{1/2} = \frac{1}{2}\left(1 \pm \sqrt{1-4M_Z^2/s}\right). \tag{B.15}$$

The expressions for the form factors I^{ZZ}, I_5^{ZZ} are valid for $s \leq 2\,\mathrm{Re}\,M^2$; those for the WW-box are obtained from (B.14, 15) by the substitution $M_Z^2 \to M_W^2$ (the gauge boson masses include as imaginary part $-iM\Gamma$).

As an illustration we present for $\sqrt{s} = 93$ GeV the functions I and I_5 for the ZZ and WW boxes in Figs. 24a, b. Only the t-channel expressions have an imaginary part comparable to the real part. In the other cases Im I is of the order of magnitude $\alpha/2\pi \cdot 10^{-2}$ or even smaller.

Fortschr. Phys. 34 (1986) 11, 753–774

First Quantum Corrections to the Classical Partition Function for a Non-Relativistic Electrodynamical System

F. Ruiz Ruiz and R. F. Alvarez-Estrada*)

Departamento de Física Teórica, Facultad de Ciencias Físicas, Universidad Complutense, 29040-Madrid, Spain

Abstract

The classical partition function for a system in thermodynamical equilibrium formed by N identical non-relativistic particles interacting through Coulomb potentials and with the dynamical electromagnetic field is studied. It is proved that the dynamical or transverse EM degrees of freedom decouple from the particle ones. It is also shown that this decoupling does to take place in the quantum mechanical partition function. The leading quantum corrections to the classical partition function are explicitly given. Such corrections are shown to be determined by instantaneous dipole-dipole coulombic interactions and by self-energy effects, and to receive no contribution from the interaction among different particles mediated by the dynamical EM field.

1. Introduction

The problem of the classical limit can be stated as that of proving that any quantum theory must tend to its classical version as the limit $\hbar \to 0$ is taken. It has no simple solution, even in the case of a particle interacting with the dynamical EM field, as Bialynicki-Birula [1] has pointed out. To treat this case coherent states have been used [1]. More explicitly, it can be proved [2] that by means of a unitary transformation, related to coherency in some sense, the classical energy is the leading, $\hbar^0$-term, of the quantum energy. Among the various contributions to the classical limit problem in Quantum Electrodynamics (QED) we shall recall those of Dente [3], Stapp [4], Halpern and Siegel [5], and Rohrlich [6] and references therein.

From the Statistical Mechanics viewpoint, the problem of the classical limit is still more interesting for two reasons. Firstly, because it is related to QED (or, even, to the more general problem of quantum field theories) at finite temperature [7, 8], so important is cosmological problems. And, secondly, because it corresponds to the much more physical situation of many charged particles.

The methods to be used in the study of Statistical Mechanics for QED consist in the evaluation of the partition function, namely, a trace. Each one of the matrix elements entering in the trace is evaluated by means of the path integral formalism. For an exposition of this formalism see Feynman and Hibbs [9, 10]. A more mathematical presentation is given by Simon [11] and Thirring [12]. This paper is devoted to the analysis of the classical limit problem for many particle electrodynamical systems and essential differences are found between the classical and quantum statistics.

*) Address for the academic year 1985–86: High Energy Physics Group, Lawrence Berkeley Laboratory, Berkeley University, California, USA.

It should be noticed that the Statistical Mechanics of and electrodynamical system is closely related to the physics of a plasma. Standard methods to study the latter, discussed by Balescu [13] or Ichimaru [14], could have been used. But for the study of the classical limit and the related problems treated in this paper the path integral formalism turns out to be more useful.

In section 2, we study a classical system in thermodynamical equilibrium formed by identical particles interacting with the dynamical EM field. It is proved that the transverse or dynamical degrees of freedom of the EM field decouple from the particle ones. The consequences of this fact on the mean values for the EM potential $\vec{A}$, the physical electric $\vec{E}$ and magnetic $\vec{B}$ fields and for the energy are derived. In section 3, the system of section 2 is studied quantum-mechanically by means of Feynman path integrals. As an aside (appendix A), we also find a bound for the quantum partition function for one particle interacting with the dynamical EM field. Section 4 is devoted to prove that when the limit is taken in the quantum partition function, the classical one is recovered. We also undertake the much more difficult task of evaluating the leading quantum corrections. Explicit expression for them are given, which constitute the main result of this work and, to our knowledge, are new. In section 5 another method of evaluating the classical limit is studied, namely, the one called quasiconstant field approximation.

2. Classical Statistical Mechanics for Non-Relativistic Electrodynamics

We consider N classical (Cl) identical charged particles of charge q and mass m interacting with the classical EM field $\vec{A}_{\text{Cl}}(\vec{x}, t)$. To avoid any stability problem we assume that there is a homogeneous background with charge of opposite sign to that of the particles and without any dynamical degrees of freedom, so that the system is electrically neutral. The Hamiltonian in the Coulomb gauge $\left(\vec{\nabla}\vec{A}_{\text{Cl}}(\vec{x}, t) = 0\right)$ is given by

$$H_{\text{Cl}} = H_{p,\text{Cl}} + H_{\text{EM,Cl}} + H_c, \quad H_{p,\text{Cl}} = \frac{1}{2m} \sum_{i=1}^{N} \left[\vec{p}_i - \frac{q}{c} \vec{A}_{\text{Cl}}(\vec{x}_i, t)\right]^2 \tag{2.1}$$

$$H_{\text{EM,Cl}} = \frac{1}{2} \int d^3\vec{x} \left[\frac{1}{c^2} \left(\frac{\partial \vec{A}_{\text{Cl}}}{\partial t}\right)^2 + (\vec{\nabla} \wedge \vec{A}_{\text{Cl}})^2\right], \quad H_c = \sum_{i>j=1}^{N} \frac{q^2}{4\pi |\vec{x}_i - \vec{x}_j|}, \tag{2.2}$$

where $\vec{x}_i$ and $\vec{p}_i$ are the position and momentum of the i th particle and the integration over $\vec{x}$ is extended to the volume V ocuppied by the system. Notice that the electrostatic (Coulomb) self-energy U_c for each particle will be discarded both in the classical and quantum cases. Their inclusion would amount to an overall constant factor $\exp(-\beta N U_c)$ in the partition function. In a linear polarization basis and as $V \to \infty$, $\vec{A}_{\text{Cl}}(\vec{x}, t)$ takes the form

$$\vec{A}_{\text{Cl}}(\vec{x}, t) = \sum_{\lambda=1,2} \int \frac{d^3\vec{k}}{(2\pi)^{3/2}} [c(\vec{k})\, \vec{\varepsilon}(\vec{k}, \lambda)\, a(\vec{k}, \lambda, t)\, e^{i\vec{k}\vec{x}} + \text{c.c.}], \tag{2.3}$$

where we have introduced the classical complex amplitudes $a(\vec{k}, \lambda, t)$ $(\lambda = 1, 2)$ and an ultraviolet cut-off factor $c(\vec{k})$, $\left(c(\vec{k}) = c^*(-\vec{k})\right)$.

Using (complex) oscillator coordinates and momenta $q(\vec{k}, \lambda, t)$ and $\pi(k, \lambda, t)$ for each mode $(\vec{k}, \lambda)$ [15]

$$\begin{aligned} q(\vec{k}, \lambda, t) &= \frac{i}{c} [a^*(-\vec{k}, \lambda, t) - a(\vec{k}, \lambda, t)], \\ \pi(\vec{k}, \lambda, t) &= -\frac{\omega}{c} [a^*(\vec{k}, \lambda, t) + a(-\vec{k}, \lambda, t)], \end{aligned} \tag{2.4}$$

with $q^*(\vec{k}, \lambda, t) = q(-\vec{k}, \lambda, t)$ and $\pi^*(\vec{k}, \lambda, t) = \pi(-\vec{k}, \lambda, t)$, the first Eq. (2.2) and Eq. (2.3) become

$$H_{\mathrm{EM,Cl}} = \frac{1}{2} \sum_{\lambda=1,2} \int \frac{d^3\vec{k}}{(2\pi)^3} |c(\vec{k})|^2 [|\pi(\vec{k}, \lambda, t)|^2 + \omega^2 |q(\vec{k}, \lambda, t)|^2], \tag{2.5}$$

$$\vec{A}_{\mathrm{Cl}}(\vec{x}, t) = ic \sum_{\lambda=1,2} \int \frac{d^3\vec{k}}{(2\pi)^{3/2}} c(\vec{k})\, \vec{\varepsilon}(\vec{k}, \lambda)\, q(\vec{k}, \lambda, t)\, e^{i\vec{k}\vec{x}}. \tag{2.6}$$

The classical partition function in the canonical ensemble reads ([13, 16])

$$Z_{N,\mathrm{Cl}} = \frac{1}{N!} \int \prod_{i=1}^{N} \frac{d^3\vec{x}_i\, d^3\vec{p}_i}{(2\pi\hbar)^3} \prod_{\vec{k},\lambda} \frac{dq(\vec{k}, \lambda)\, d\pi(\vec{k}, \lambda)}{2\pi\hbar} \exp(-\beta H_{\mathrm{Cl}}) \tag{2.7}$$

with H_{Cl} given by Eqs. (2.1)–(2.2) and (2.5)–(2.6). Notice that: i) all oscillator coordinates and momenta, for any $(\vec{k}, \lambda)$, are taken at a given $t = 0$ (not written explicitly), and ii) the right-hand-side of Eq. (2.7) contains a functional integration over $q(\vec{k}, \lambda)$ and $\pi(\vec{k}, \lambda)$.

Next we change $\vec{p}_i - q/c\, \vec{A}_{\mathrm{Cl}}(\vec{x}_i, t) \to \vec{p}_i$ in (2.7). Then, we perform successive gaussian integrations, first, over all $\vec{p}_i$ and, second, over all q's and π's. Thus, we get

$$Z_{N,\mathrm{Cl}} = \frac{1}{N!} \left(\frac{m}{2\pi\beta\hbar^2}\right)^{3N/2} Z_{\mathrm{EM,Cl}}(\beta, \hbar) \int \prod_{i=1}^{N} d^3\vec{x}_i \cdot \exp\left\{-\beta \sum_{i>j=1}^{N} \frac{q^2}{4\pi |\vec{x}_i - \vec{x}_j|}\right\}. \tag{2.8}$$

where $Z_{\mathrm{EM,Cl}}$ is the partition function for the classical EM field, namely:

$$Z_{\mathrm{EM,Cl}}(\beta, \hbar) = \int \prod_{\vec{k},\lambda} \frac{dq(\vec{k}, \lambda)\, d\pi(\vec{k}, \lambda)}{2\pi\hbar} \exp(-\beta H_{\mathrm{EM,Cl}}) = \prod_{\vec{k}} \left(\frac{1}{\omega^2\beta^2\hbar^2}\right). \tag{2.9}$$

Therefore, as far as the partition function is concerned, things occur as if the particles interact only through the instantaneous Coulomb potential while the transverse EM field decouples from the former and behaves as a free field. This fact, to our knowledge, does not seem to have been pointed out before. It also seems to be a direct consequence of the minimal coupling (gauge invariance). Notice that Eq. (2.9) yields the Rayleigh-Jeans law. In fact, by taking logarithms in Eq. (2.9) we obtain the energy per unite volume as

$$\frac{1}{V} \langle H_{\mathrm{EM,Cl}} \rangle = -\frac{1}{v} \frac{\partial}{\partial\beta} \ln Z_{\mathrm{EM,Cl}}(\beta, \hbar) = \sum_{\lambda=1,2} \int \frac{d^3\vec{k}}{(2\pi)^3} \frac{1}{\beta}. \tag{2.10}$$

In turn, by using $\sum_{\lambda=1,2} d^3\vec{k}(2\pi)^{-3} = 8\pi\nu^2 c^{-3}\, d\nu$ in Eq. (2.10), we get the energy per unit volume between ν and $\nu + d\nu$ as $8\pi\nu^2(c^3\beta)^{-1}$, which is the Rayleigh-Jeans formula.

Taking the partition function as starting point it is easy to evaluate the classical mean value of any dynamical variable. For instance, since $\vec{A}_{\mathrm{Cl}}(\vec{x}, t)$ is linear in $q(\vec{k}, \lambda)$ we have, by symmetrical integration

$$\langle \vec{A}_{\mathrm{Cl}}(\vec{x}) \rangle = \frac{1}{Z_{\mathrm{N,Cl}}} \frac{1}{N!} \int \prod_{c=1}^{N} \frac{d^3\vec{x}_i\, d^3\vec{p}_i}{(2\pi\hbar)^3} \prod_{\vec{k},\lambda} \frac{dq(\vec{k}, \lambda)\, d\pi(\vec{k}, \lambda)}{2\pi\hbar} \vec{A}_{\mathrm{Cl}}(\vec{x}) \exp(-\beta H_{\mathrm{Cl}}) = 0. \tag{2.11}$$

Analogously, for the magnetic field $\vec{B}_{\mathrm{Cl}} = \vec{\nabla} \wedge \vec{A}_{\mathrm{Cl}}$ we get

$$\langle \vec{B}_{\mathrm{Cl}}(\vec{x}) \rangle = 0. \tag{2.12}$$

For the electric field $\vec{E} = -(\partial \vec{A}_{\rm Cl}/c\,\partial t) + \vec{\nabla} \sum_{i=1}^{N} (q/4\pi\,|\vec{x} - \vec{x}_i|)$ we have $\langle \vec{E}_{\rm Cl}(\vec{x})\rangle = \vec{\nabla}\langle V_{\rm Cl}(\vec{x})\rangle$, with

$$\langle V_{\rm Cl}(\vec{x})\rangle = \frac{1}{Z_{N,\rm Cl}}\frac{1}{N!}\int \prod_{i=1}^{N} \frac{d^3\vec{x}_i\, d^3\vec{p}_i}{(2\pi\hbar)^3} \prod_{\vec{k},\lambda} \frac{dq(\vec{k},\lambda)\, d\pi(\vec{k},\lambda)}{2\pi\hbar} \sum_{i=1}^{N} \frac{q}{4\pi\,|\vec{x} - \vec{x}_i|} \exp(-\beta H_{\rm Cl}),$$

since in the Coulomb gauge $-\partial \vec{A}_{\rm Cl}/c\,\partial t$ is the canonical momentum,

$$-\frac{1}{c}\frac{\partial A_{\rm Cl}(\vec{x},t)}{\partial t} = -i\sum_{\lambda=1,2}\int \frac{d^3\vec{k}}{(2\pi)^{3/2}}\, c(\vec{k})\, \vec{\varepsilon}(\vec{k},\lambda)\, \pi(\vec{k},\lambda)\, e^{i\vec{k}\vec{x}},$$

and, by symmetrical integration in $\pi(\vec{k},\lambda)$ we get

$$\left\langle \frac{\partial \vec{A}_{\rm Cl}(\vec{x},t)}{\partial t} \right\rangle = \frac{\langle \partial \vec{A}_{\rm Cl}(\vec{x})\rangle}{\partial t} = 0.$$

For any vector $\vec{a}$, by performing an overall translation $\vec{x}_i \to \vec{x}_i + \vec{a}$, we get $\langle V_{\rm Cl}(\vec{x} + \vec{a})\rangle = \langle V_{\rm Cl}(\vec{x})\rangle$, or, equivalently, $\langle V_{\rm Cl}(\vec{x})\rangle = \text{const.}$, and therefore,

$$\langle \vec{E}_{\rm Cl}(\vec{x})\rangle = 0. \tag{2.13}$$

By a straightforward generalization of the latter arguments it follows that if the background is not homogeneous but a periodic distribution of heavy charges without internal degrees of freedom, the mean value for the electrostatic potential and for the electric field are periodic with the same periodicity as that of the background.

For the kinetic part $H_{p,\rm Cl}$ of the hamiltonian, by performing the gaussian integrations, we have

$$\langle H_{p,\rm Cl}\rangle = \frac{1}{Z_{N,\rm Cl}}\frac{1}{N!}\int \prod_{i=1}^{N} \frac{d^3\vec{x}_i\, d^3\vec{p}_i}{(2\pi\hbar)^3} \prod_{\vec{k},\lambda} \frac{dq(\vec{k},\lambda)\, d\pi(\vec{k},\lambda)}{2\pi\hbar} H_{p,\rm Cl} \exp(-\beta H_{\rm Cl})$$

$$= \frac{3N}{2}\frac{1}{\beta}. \tag{2.14}$$

And for the electrostatic energy $H_{\rm C}$ we get

$$\langle H_{c,\rm Cl}\rangle = \left\{\int \prod_{i=1}^{N} d^3\vec{x}_i \exp(-\beta H_c)\right\}^{-1} \int \prod_{i=1}^{N} d^3x_i H_c \exp(-\beta H_c). \tag{2.15}$$

From Eqs. (2.1)–(2.2), (2.10) and (2.14)–(2.15) it follows

$$\langle H_{\rm Cl}\rangle = \frac{3N}{2}\frac{1}{\beta} + V\sum_{\lambda=1,2}\int \frac{d^3\vec{k}}{(2\pi)^3}\frac{1}{\beta} + \langle H_{c,\rm Cl}\rangle = -\frac{\partial}{\partial\beta}\ln Z_{N,\rm Cl}. \tag{2.16}$$

3. Quantum Statistical Mechanics for Non-Relativistic Electrodynamics

Consider N quantum (Q) identical particles (bosons or fermions) interacting with a quantized EM field $\vec{A}_Q(\vec{x})$. The hamiltonian H_Q for the system in the Coulomb gauge $\left(\vec{\nabla}\vec{A}_Q(\vec{x}) = 0\right)$ and in the Schrodinger picture is given by

$$H_Q = H_{p,Q} + H_{{\rm EM},Q} + H_c, \quad H_{p,Q} = \frac{1}{2m}\sum_{c=1}^{N}\left[\vec{P}_i - \frac{q}{c}\vec{A}_Q(\vec{x}_i)\right]^2, \tag{3.1}$$

$$H_{\mathrm{EM},Q} = \frac{1}{2} \sum_{\lambda=1,2} \int \frac{d^3\vec{k}}{(2\pi)^3} |c(\vec{k})|^2 [|\Pi(\vec{k},\lambda)|^2 + \omega^2 |Q(\vec{k},\lambda)|^2],$$
$$H_c = \sum_{i>j=1}^{N} \frac{q^2}{4\pi |\vec{x}_i - \vec{x}_j|}, \tag{3.2}$$

where

$$\vec{A}_Q(\vec{x}) = ic \sum_{\lambda=1,2} \int \frac{d^3\vec{k}}{(2\pi)^{3/2}} c(\vec{k})\, \vec{\varepsilon}(\vec{k},\lambda)\, Q(\vec{k},\lambda)\, e^{i\vec{k}\vec{x}}. \tag{3.3}$$

$Q(\vec{k},\lambda)$ and $\Pi(\vec{k},\lambda)$ are the operators associated to $q(\vec{k},\lambda,t)$ and $\pi(\vec{k},\lambda,t)$. They satisfy canonical commutation rules and

$$Q^+(\vec{k},\lambda) = Q(-\vec{k},\lambda), \qquad \Pi^+(\vec{k},\lambda) = \Pi(-\vec{k},\lambda). \tag{3.4}$$

Eq. (3.4) tells us that even if $\vec{A}_Q(\vec{x})$ is selfadjoint, $Q(\vec{k},\lambda)$ and $\Pi(\vec{k},\lambda)$ are not. In the first Eq. (3.2) we have omitted the energy of the vacuum. Later on, we will come back to this point.

In the subspace of N identical particles in an EM field, the set

$$\left\{\frac{1}{\sqrt{N!}} \sum_P (\mp 1)^{\varepsilon(P)} |\vec{x}_1(P1), \ldots, \vec{x}_N(PN); \{\vec{q}\}\rangle\right\}$$

forms an orthonormal basis. The sum $\sum_P$ is extended to all possible different permutations of N objects. The minus sign is for fermions and the plus one for bosons while $\varepsilon(P)$ is the order of each permutation P. The state $|\vec{x}_1(P1), \ldots, \vec{x}_N(PN); \{\vec{q}\}\rangle$ is that one in which particle 1 is in $\vec{x}_{P1}, \ldots$, particle N in $\vec{x}_{PN}$ and the harmonic oscillators of the EM field are in a state denoted by $\{q\}$. Using this basis, the quantum partition function in the canonical ensemble is given by (compare with [10], section 10.4)

$$\begin{aligned} Z_{N,Q} &= \mathrm{Tr}\,[\exp(-\beta H_Q)] \\ &= \frac{1}{N!} \int \prod_{i=1}^{N} d^3\vec{x}_i \prod_{\vec{k},\lambda} d\vec{q}(\vec{k},\lambda) \sum_P (\mp 1)^{\varepsilon(P)} \langle \vec{x}_1, \ldots, \vec{x}_N; \{\vec{q}\} \\ &\quad \times |\exp(-\beta H_Q)\, |\vec{x}_1(P1), \ldots, \vec{x}_N(\mathrm{PN}); \{\vec{q}\}\rangle. \end{aligned} \tag{3.5}$$

All the matrix elements appearing in Eq. (3.5) are "probability amplitudes" for the system being at $\tau = 0$ in $|\vec{x}_1(P1), \ldots, \vec{x}_N(PN); \{\vec{q}\}\rangle$ to be at $\tau = \beta\hbar$ in $|\vec{x}_1, \ldots, \vec{x}_N; \{\vec{q}\}\rangle$, which can be represented by means of Feynman path integrals as

$$\begin{aligned} &\langle \vec{x}_1, \ldots, \vec{x}_N; \{\vec{q}\}|\exp(-\beta H_Q)\,|\,\vec{x}_1', \ldots, \vec{x}_N; \{\vec{q}\}\rangle \\ &= \int \prod_{i=1}^{N} \prod_{\tau} \frac{d^3\vec{x}_i(\tau)\, d^3\vec{p}_i(\tau)}{(2\pi\hbar)^3} \prod_{\vec{k},\lambda,\tau} \frac{dq(\vec{k},\lambda,\tau)\, d\pi(\vec{k},\lambda,\tau)}{2\pi\hbar} \\ &\quad \times \exp\left\{\frac{1}{\hbar} \int_0^{\beta\hbar} d\tau \left[i \sum_{j=1}^{N} \vec{p}_j(\tau)\, \dot{\vec{x}}_j(\tau) + i \sum_{\lambda=1,2} \int d^3\vec{k}\pi(\vec{k},\lambda,\tau)\, \dot{q}(\vec{k},\lambda,\tau) - H_{\mathrm{Cl}}(\tau)\right]\right\}, \end{aligned} \tag{3.6}$$

where $H_{\mathrm{Cl}}(\tau)$ is given in section 2. Evaluating Eq. (3.6) (see appendix A) and substituting it into (3.5) we get

$$Z_{N,Q} = \frac{1}{N!} Z_{\mathrm{EM},Q}(\beta, \hbar) \int \prod_{i=1}^{N} d^3\vec{x}_i \sum_{P} (\mp 1)^{\varepsilon(P)} \prod_{j=1}^{N} \int_{\vec{x}_j(P_j)}^{\vec{x}_j} D^3\vec{y}_j \exp\left[G(\vec{y})\right] \tag{3.7}$$

$$Z_{\mathrm{EM},Q}(\beta, \hbar) = \prod_{\vec{k}} \frac{1}{2(\cosh \omega\beta\hbar - 1)}, \tag{3.8}$$

where we are using the compact notation

$$D^3\vec{y}_i \equiv \lim_{M\to\infty} \left(\frac{m}{2\pi\hbar\delta}\right)^{3/2} \prod_{l=1}^{M} \frac{d^3\vec{y}_{i,l}}{(m^{-1}2\pi\hbar\delta)^{3/2}} \tag{3.9}$$

and the exponent $G(\vec{y})$ is given by

$$G(\vec{y}) \equiv \frac{1}{\hbar} \int_0^{\beta\hbar} d\tau \left[-\frac{m}{2} \sum_{i=1}^{N} \dot{\vec{y}}_i^{\,2}(\tau) - \sum_{i>j=1}^{N} \frac{q^2}{4\pi\,|\vec{y}_i(\tau) - \vec{y}_j(\tau)|} \right]$$
$$- \frac{q^2}{2\hbar} \sum_{i,j=1}^{N} \sum_{\lambda=1,2} \int \frac{d^3\vec{k}}{(2\pi)^3} \frac{|c(\vec{k})|^2}{\omega} \int_0^{\beta\hbar} d\tau \int_0^{\beta\hbar} d\tau' [\vec{\varepsilon}(\vec{k}, \lambda)\, \dot{\vec{y}}_i(\tau)]$$
$$\times \exp\{i\vec{k}[\vec{y}_i(\tau) - \vec{y}_j(\tau)]\}\, f(\tau - \tau', \beta\hbar)\, [\vec{\varepsilon}(\vec{k}, \lambda)\, \dot{\vec{y}}_j(\tau')]. \tag{3.10}$$

The function $f(\tau - \tau, \beta\hbar)$ has the form

$$f(\tau - \tau', \beta\hbar) = \frac{e^{-\omega|\tau-\tau'|} + e^{\omega|\tau-\tau'|}\, e^{\omega\beta\hbar}}{1 - e^{-\omega\beta\hbar}}. \tag{3.11}$$

Remember that for a partition of the interval $[0, \beta\hbar]$ into $M + 1$ subintervals of length δ, and according to the midpoint rule [17], the factor $\exp\{i\vec{k}(\vec{y}_i(\tau) - \vec{y}_j(\tau'))\}$ gives rise to

$$\exp\left\{\frac{i}{2}\, \vec{k}[(\vec{y}_{i,l} + \vec{y}_{i,l-1}) - (\vec{y}_{j,l'} + \vec{y}_{j,l'-1})]\right\}.$$

The partition function for the EM field $Z_{\mathrm{EM},Q}(\beta, \hbar)$, Eq. (3.8), can be written as

$$Z_{\mathrm{EM},Q}(\beta, \hbar) = \prod_{\vec{k},\lambda} \frac{e^{-\omega\beta\hbar/2}}{1 - e^{-\omega\beta\hbar}}. \tag{3.12}$$

The factor $\exp(-\omega\beta\hbar/2)$ corresponds to the ground state (vacuum state in a Fock space) of the oscillator associated to $(\vec{k}, \lambda)$. In all what we have done we have ignored the zero-point energy shift in such a way that instead of the $H_{\mathrm{EM},Q}$ of Eq. (3.2) we should have written

$$H_{\mathrm{EM},Q} - \frac{\hbar}{2} V \sum_{\lambda=1,2} \int \frac{d^3\vec{k}}{(2\pi)^3} |c(\vec{k})|^2\, \omega\,.$$

This extra term cancels the contribution from the ground state in $Z_{\mathrm{EM},Q}(\beta, \hbar)$ and transforms Eq. (3.12) into

$$Z_{\mathrm{EM},Q}(\beta, \hbar) = \prod_{\vec{k},\lambda} \frac{1}{1 - e^{-\omega\beta\hbar}},$$

which is Planck's partition function. See, for instance, [15] for a study of the partition function for the free quantized EM field.

Notice that Eq. (3.7) is the product of the quantum partition functions for the free EM field and for N particles interacting instantaneously through a Coulomb potential and at different times by means of a "propagator"

$$D_{ij}^{-1}(\vec{y} - \vec{y}', \tau - \tau') \equiv \sum_{\lambda=1,2} \int \frac{d^3\vec{k}}{(2\pi)^3} \frac{|c(\vec{k})|^2}{\omega} \varepsilon_i(\vec{k}, \lambda) \times \exp\{i\vec{k}[\vec{y}(\tau) - \vec{y}'(\tau')]\} f(\tau - \tau', \beta\hbar) \varepsilon_j(k, \lambda). \quad (3.13)$$

We also notice that Eq. (3.7) accomplishes FEYNMAN's elimination of the EM field variables with full generality in the partition function: compare with [10], chapter 9 and 11.

Next, we evaluate the averages values of the quantized operators $\vec{A}_Q(\vec{x})$, $\vec{B}_Q(\vec{x})$, $\vec{E}_Q(\vec{x})$ and H_Q. For any operator 0 we have

$$\langle 0 \rangle = \frac{\mathrm{Tr}\,[\exp(-\beta H_Q)\, 0]}{Z_{N,Q}}.$$

In the case of $\langle \vec{A}_Q(\vec{x}) \rangle$ the matrix elements needed to evaluate the trace are of the type

$$\langle \vec{x}_1, \ldots, \vec{x}_N; \{\vec{q}\}| \exp(-\beta H_Q)\, Q(\vec{k}, \lambda)\, |\vec{x}_1(P1), \ldots, \vec{x}_N(PN); \{\vec{q}\}\rangle$$
$$= \langle \vec{x}_1, \ldots, \vec{x}_N; \{\vec{q}\}| \exp(-\beta H_Q)\, |\vec{x}_1(P1), \ldots, \vec{x}_N(PN); \{\vec{q}\}\rangle\, \vec{q}(\vec{k}, \lambda),$$

so, proceeding in the same way as in the partition function, after a gaussian integration, we get

$$\langle \vec{A}_Q(\vec{x}) \rangle = \frac{Z_{\mathrm{EM},Q}(\beta, \hbar)}{N!\, Z_{N,Q}} \sum_{\lambda=1,2} \int d^3\vec{k}\, \vec{\varepsilon}(\vec{k}, \lambda)\, e^{i\vec{k}\vec{x}} \int \prod_{i=1}^{N} d^3\vec{x}_i \sum_P (\mp 1)^{\varepsilon(P)} \times \prod_{j=1}^{N} \int_{\vec{x}_j(P_j)}^{\vec{x}_j} D^3\vec{y}_j \frac{(C + D)\,(e^{\omega\beta\hbar} - 1)}{(\cosh \omega\beta\hbar - 1)}\, e^{G(\vec{y})}, \quad (3.14)$$

with C and D given by

$$C = -\frac{1}{2\omega} \int_0^{\beta\hbar} d\tau e^{-\omega\tau} \sum_{i=1}^{N} \gamma_i(\tau), \qquad D = -\frac{1}{2\omega} \int_0^{\beta\hbar} d\tau e^{-\omega(\beta\hbar - \tau)} \sum_{i=1}^{N} \gamma_i(\tau), \quad (3.15)$$

$$\gamma_i(\tau) = q \frac{c(\vec{k})}{(2\pi)^{3/2}}\, \vec{\varepsilon}(\vec{k}, \lambda)\, \dot{\vec{y}}_i(\tau)\, e^{i\vec{k}\vec{y}_i(\tau)}. \quad (3.16)$$

Analogously, we get

$$\langle \vec{B}_Q(\vec{x}) \rangle = \frac{Z_{\mathrm{EM},Q}(\beta, \hbar)}{N!\, Z_{N,Q}} \sum_{\lambda=1,2} \int d^3\vec{k}\vec{k} \wedge \vec{\varepsilon}(\vec{k}, \lambda)\, e^{i\vec{k}\vec{x}} \int \prod_{i=1}^{N} d^3\vec{x}_i \sum_P (\mp 1)^{\varepsilon(P)} \times \prod_{j=1}^{N} \int_{\vec{x}_j(P_j)}^{\vec{x}_j} D^3\vec{y}_j \frac{(C + D)\,(e^{\omega\beta\hbar} - 1)}{(\cosh \omega\beta\hbar - 1)}\, e^{G(\vec{y})}. \quad (3.17)$$

The electric field is the sum of two terms,

$$\langle \vec{E}_Q(\vec{x}) \rangle = \left\langle i \sum_{\lambda=1,2} \int \frac{d^3\vec{k}}{(2\pi)^{3/2}}\, c(\vec{k})\, \vec{\varepsilon}(\vec{k}, \lambda)\, \Pi(\vec{k}, \lambda)\, e^{i\vec{k}\vec{x}} \right\rangle + \left\langle \vec{\nabla} \sum_{i=1}^{N} \frac{q}{4\pi\, |\vec{x} - \vec{x}_i|} \right\rangle .$$

In the first one we have matrix elements of type

$$\langle \vec{x}_1, \ldots, \vec{x}_N; \{\vec{q}\}|\, \exp(-\beta H_Q)\, \Pi(\vec{k}, \lambda)\, |\vec{x}_1(P1), \ldots, \vec{x}_N(PN); \{\vec{q}\}\rangle .$$

To evaluate them we first perform the integration over all the $\pi(\vec{k}, \lambda)$, which is gaussian, and get a factor

$$\int_0^{\beta\hbar} d\tau \dot{q}(\vec{k}, \lambda, \tau),$$

which, by the trace condition $q(\vec{k}, \lambda, \beta\hbar) = q(\vec{k}, \lambda, 0) = \tilde{q}(\vec{k}, \lambda)$, is identically zero. Therefore,

$$\langle \vec{E}_Q(\vec{x}) \rangle = \frac{Z_{\mathrm{EM},Q}(\beta, \hbar)}{N!\, Z_{N,Q}} \int \prod_{i=1}^{N} d^3\vec{x}_i \left(\vec{\nabla} \sum_{i=1}^{N} \frac{q}{4\pi\, |\vec{x} - \vec{x}_i|} \right)$$
$$\times \sum_P (\mp 1)^{\varepsilon(P)} \prod_{j=1}^{N} \int_{\vec{x}_j(P_j)}^{\vec{x}_j} D^3\vec{y}_j e^{G(\vec{y})} . \tag{3.18}$$

In the quantum case, and differently from what happens in the classical one, the integration of the EM variables gives rise to an interaction among the particles, bilinear in their velocities and propagated by the function given in Eq. (3.13). Besides, $\langle \vec{A}_Q(\vec{x}) \rangle$ and $\langle \vec{B}_Q(\vec{x}) \rangle$ contain the sum $C + D$, which is basically a velocity, see Eqs. (3.14)—(3.17).

4. Classical Limit and Leading Quantum Corrections for the Partition Function

In this section we shall study what happens with the quantum quantities when the classical limit $\hbar \to 0$ is taken at fixed non-vanishing temperature ($mc^2 \gg \beta^{-1} \neq 0$). We start with the partition function, Eq. (3.7). To analyse the behaviour of the path integral of (3.7) for $\hbar \to 0$ we assume that the trajectories $\vec{y}_i(\tau)$ can be written in general as

$$\vec{y}_i(\tau) = \vec{x}_i(Pi) + \frac{\vec{x}_i - \vec{x}_i(Pi)}{\beta\hbar}\, \tau + \vec{\xi}_i(\tau) \tag{4.1}$$

with $\vec{\xi}_i(\tau)$ satisfying the boundary conditions

$$\vec{\xi}_i(0) = \vec{\xi}_i(\beta\hbar) = 0 . \tag{4.2}$$

We recall (see, for instance, [16], chapter 10) that for N noninteracting identical particles the partition function is given by

$$Z^0_{N,Q} = \frac{1}{N!} \left(\frac{m}{2\pi\beta\hbar^2} \right)^{3N/2} \sum_P (\mp 1)^{\varepsilon(P)} \int \prod_{i=1}^{N} d^3\vec{x}_i \exp\left\{ -\frac{m}{2\beta\hbar^2} \sum_{i=1}^{N} [\vec{x}_i - \vec{x}_i(Pi)]^2 \right\},$$

so that if we let $\hbar \to 0$ the only permutation which contributes is the identity (I):

$$Z^0_{N,Q} \xrightarrow[\hbar \to 0]{} \frac{1}{N!} \left(\frac{m}{2\pi\beta\hbar^2} \right)^{3N/2} V^N = Z^0_{N,\mathrm{Cl}},$$

V being the volume occupied by the particles. Moreover, when the particles interact through an instantaneous local potential, the leading quantum corrections to the free energy of orders $\hbar$ and $\hbar^2$ (at fixed $\beta \neq \infty$) are also due to the identity. The contributions arising from quantum statistics (say, other permutations different from 1) produce exchange corrections of higher order in $\hbar$ both for non-interacting particles and for particles interacting through instantaneous potentials: their specific form (of order $\hbar^3$) for the free energy in the case of non-interacting particles is well known. About these matters, see, for instance, chapters 10 and 11 in [16] and section 33 in [18]. This tells us that when we introduce, in addition, the interaction mediated by the dynamical EM field which, for several purposes, could be regarded as a small perturbation, the leading quantum corrections to the partition function (to be summarized neatly by Eq. (4.5), our main result) will also come from the identity permutation. The exchange corrections due to other permutations different from I will be of higher order in $\hbar$ in the present case as well, (see also section 5). Accordingly, in Eq. (3.7) we will take $P = I$, which implies that Eq. (4.1) should be replaced by

$$\vec{y}_i(\tau) = \vec{x}_i + \vec{\xi}_i(\tau), \qquad \vec{\xi}_i^{\,2}(\tau) \sim \frac{\beta\hbar^2}{m}. \tag{4.3}$$

The last estimate of the size of $\vec{\xi}_i(\tau)$ comes from equaling the kinetic energy to 1 and agrees with the standard philosophy for brownian paths, [17]. Notice that

$$\dot{\vec{\xi}}_i^{\,2}(\tau) \sim \frac{1}{\beta m} \tag{4.4}$$

and that $\tau, \tau', |\tau - \tau'| \sim \beta\hbar$. A lengthy evaluation of Eq. (3.7) under the hypothesis (4.3) and (4.4) yields our main result, (for a detailed derivation, see appendix B):

$$Z_{N,Q} = \frac{1}{N!} Z_{\mathrm{EM},Q}(\beta,\hbar) \left(\frac{m}{2\pi\beta\hbar^2}\right)^{3N/2} \int \prod_{i=1}^{N} d^3\vec{x}_i \exp\left\{-\beta \sum_{i>j=1}^{N} \frac{q^2}{4\pi|\vec{x}_i - \vec{x}_j|}\right\}$$
$$\times \left\{1 - 2N\frac{\beta\hbar^2}{m^2}\mathscr{C} + 2N\frac{\beta^2\hbar^2c^2}{m}\mathscr{C} + \frac{3}{2}\frac{\beta^3\hbar^2}{m}\left(\frac{q^2}{2\pi}\right)^2 \Delta + O(\hbar^3)\right\}, \tag{4.5}$$

with $\mathscr{C}$ given by

$$\mathscr{C} \equiv \left(\frac{q}{c}\right)^2 \int \frac{d^3\vec{k}}{(2\pi)^3} |c(\vec{k})|^2. \tag{4.6}$$

For $N = 1$, (4.5) reduces to

$$Z_{N=1,Q} = Z_{\mathrm{EM},Q}(\beta,\hbar)\left(\frac{m}{2\pi\beta\hbar^2}\right)^{3/2}\left\{1 - 2\frac{\beta\hbar^2}{m^2}\mathscr{C} + 2\frac{\beta^2\hbar^2c^2}{m}\mathscr{C} + O(\hbar^3)\right\}.$$

Notice that the last curly bracket is equal to or smaller than 1 since $\beta^{-1} \ll mc^2$, in agreement with the bound (A.7) given in appendix A.

Eq. (4.5) together with

$$Z_{\mathrm{EM},Q}(\beta,\hbar) = \prod_{\vec{k}} \frac{1}{2(\cosh\omega\beta\hbar - 1)} \xrightarrow[\hbar\to 0]{} \prod_{\vec{k}} \frac{1}{\omega^2\beta^2\hbar^2} = Z_{\mathrm{EM,Cl}}(\beta,\hbar) \tag{4.7}$$

implies

$$Z_{N,Q} \xrightarrow[\hbar\to 0]{} Z_{N,\mathrm{Cl}}, \tag{4.8}$$

which is the generalization of theorem 15.9 of [11] to dynamical EM fields. Notice also that Eq. (4.5) gives explicit expressions for the first quantum corrections to the classical partition function, and, therefore, to the classical behaviour of N identical charged particles in an EM field.

Next, we study the limit $\hbar \to 0$ of the average values of $\vec{A}_Q$, $\vec{B}_Q$ and $\vec{E}_Q$. Proceeding in the same way as for $Z_{N,Q}$ (see appendix C for details) we get

$$\langle \vec{A}_Q(\vec{x}) \rangle \xrightarrow[\hbar \to 0]{} 0 = \langle \vec{A}_{\text{Cl}}(\vec{x}) \rangle, \tag{4.9}$$

$$\langle \vec{B}_Q(\vec{x}) \rangle \xrightarrow[\hbar \to 0]{} 0 = \langle \vec{B}_{\text{Cl}}(\vec{x}) \rangle, \tag{4.10}$$

$$\langle \vec{E}_Q(\vec{x}) \rangle \xrightarrow[\hbar \to 0]{} 0 = \langle \vec{E}_{\text{Cl}}(\vec{x}) \rangle. \tag{4.11}$$

For the hamiltonian we have, that by differentiating logarithmically with respect to β and taking the limit $\hbar \to 0$:

$$\langle H_Q \rangle \xrightarrow[\hbar \to 0]{} \langle H_{\text{Cl}} \rangle. \tag{4.12}$$

Eqs. (4.8)–(4.12) tell us that when the limit $\hbar \to 0$ is taken the classical results are recovered.

Before finishing this section same remarks are in order.

Remark 1. The method used here, based heavily upon Feynman's path integrals, seems far more advantageous for the present problem than the one inspired in the semiclassical approximation of Wigner-Kirkwood [19], where one would be led to solve some sort of Bloch equation by approximate methods. Moreover, our study is in the line of the work of Feynman [9], where he recovered the Schrödinger equation from the path integral formalism by studying the behaviour of the integral for small times. The validity of Eq. (4.8), the fact that without the EM field interaction the quantum fermion system with pure Coulomb forces coincides with the classical Coulomb system as $\hbar \to 0$, and the proof of stability of fermion matter for pure Coulomb forces by Lieb and Thirring [20] indicates that, at least, in the classical limit, stability of matter should also hold for the system described by Eq. (3.7).

Remark 2. Eqs. (4.5) and (4.6) seem to imply that when the cut-off function $c(\vec{k})$ is removed (namely, one would set $c(\vec{k}) = 1$ for $|\vec{k}| \leq k_{\max}$, $c(\vec{k}) = 0$ for $|\vec{k}| > k_{\max}$, with $k_{\max} \to +\infty$), one would find unavoidably a cubic ultraviolet divergence $\left(\mathscr{C} = (q/c)^2 \times (k_{\max}^3/6\pi^2)\right)$ in the leading quantum corrections to the parition function, even for $N = 1$. Such a cubic divergence would be higher than the linear ultraviolet divergence which appears when studying virtual-photon corrections to the electron self-mass in non-relativistic Quantum Electrodynamics at zero temperature (see, for instance, section 15b in [21]).

It turns out that the cubic ultraviolet divergence is superfluous, as the following discussion will show. In fact, the hamiltonian (3.1)–(3.3) for non-relativistic Quantum Electrodynamics and the partition function (3.5) are physically meaningful provided that: i) $\hbar k_{\max} \ll mc$, ii) $\beta^{-1} \ll mc^2$. Moreover, the calculation of the quantum corrections to the partition function carried out in this section has physical sense at suitably high temperatures. All these suggest that β^{-1} could vary in some interval $|(\hbar k_{\max})^2/2m| - \varepsilon_1 < \beta^{-1} < |(\hbar k_{\max})^2/2m| + \varepsilon_2$, where ε_1 should be small compared to $(\hbar k_{\max})^2/2m$ and ε_2 should be such that the above condition ii) be fulfilled.

For a quick estimate, let β^{-1} be of order $(\hbar k_{\max})^2/2m$, which is consistent with both i) and ii). Then, the two leading quantum corrections in Eq. (4.5) become

$$\frac{\beta \hbar^2}{m} \mathscr{C} \simeq \frac{1}{3\pi^2} \frac{q^2}{\hbar c} \frac{\hbar k_{\max} c}{mc^2}, \quad \frac{\beta^2 \hbar^2 c^2}{m} \mathscr{C} \simeq \frac{\beta mc^2}{3\pi^2} \frac{q^2}{\hbar c} \frac{\hbar k_{\max} c}{mc^2},$$

that is, the cubic ultraviolet divergence has disappeared and become linear. One arrives, clearly, at the same conclusion if β^{-1} varies in the range $|(\hbar k_{\max})^2/2m| - \varepsilon_1 < \beta^{-1} < |(\hbar k_{\max})^2/2m| + \varepsilon_2$. In a forthcoming paper [22] this linear divergence is removed using Quantum Field Theory techniques.

Remark 3. Let us describe the problem in a simple way using its characteristic lengths. The first one is the inteparticle distance d related to the density through $\varrho = d^{-3}$. Other three characteristic lengths are the thermal λ_{th}, de Broglie λ_B and Compton λ_C wavelengths, given by

$$\lambda_{\text{th}} = \left(\frac{m}{2\pi\beta\hbar^2}\right)^{-1/2}, \qquad \lambda_B = \frac{\hbar}{(2mE)^{1/2}}, \qquad \lambda_C = \frac{\hbar}{mc}.$$

For the actual non-relativistic system (recall conditions i) and ii) of remark 2) the following relation holds: $\lambda_{\text{th}} \gg \lambda_C$. Moreover, the wavelength λ of a photon is such that $2\pi/\lambda = |\vec{k}| < k_{\max}$, so $\lambda \gg \lambda_C$. The condition for the classical limit (validity of the equipartition principle) amounts to $\lambda_{\text{th}} \sim \lambda_B$. Notice that for the special case $\lambda \sim \lambda_{\text{th}} \sim \lambda_B$, $|\vec{k}| \sim 2\pi/\hbar \times (m/\beta)^{1/2}$, that implies $\hbar|\vec{k}|/mc \sim (\beta mc^2)^{-1/2} \ll 1$, which is consistent (with condition i)).

When restricting ourselves to the identity permutation in the analysis of the classical limit we are using implicitly $d \gg \lambda_{\text{th}}$.

For completeness, one could add a fifth length, the Debye wavelength $\lambda_D = (q^2\varrho\beta)^{-1/2} \gg d$. For a detailed discussion see [13—14].

Remark 4. It is interesting to point out the analogies and differences between the problem studied here and the polaron one, (one electron interacting with a scalar phonon field) as studied by FEYNMAN [15]. There, the path integral formalism is also used, but restricted to very large β, which means replacing Eq. (3.11) by $\exp(-\omega|\tau - \tau'|)$. Notice that in Eq. (3.11) β appears multipled by $\hbar$, so, making the approximation $\beta \to \infty$ in Eq. (3.11) would be equivalent in some way to $\hbar \to \infty$, which differs from what we have done ($\hbar \to 0$). There are also some other differences due to the vectorial nature of the EM field, the minimal coupling, the number of particles, etc.

5. Discussion of Other Methods: The Quasiconstant Field Approximation

In this section we are going to recover the classical result for the partition function by using a different method, which may well be called the quasiconstant field approximation (compare with Halpern and Siegel [5]). We shall also discuss the advantages and disadvantages of this method.

The starting point is the partition function (3.5). Now, to evaluate the propagator (3.6) we divide the $\vec{x}$-space into small cells α of volume V_α, and in each of them we expand the EM field into a power series around the center $\vec{x}_\alpha$ of the cell:

$$\vec{A}_\alpha(\vec{x}) = \vec{A}(\vec{x}_\alpha) + (x - x_\alpha)^i\,[\partial_i\vec{A}(\vec{x})]_{\vec{x}_\alpha} + \cdots. \tag{5.1}$$

By considering only the first term in (5.1) (quasiconstant field approximation), taking into account that in this approximation $\vec{P}$ and $\vec{A}_\alpha(\vec{x})$ conmute, proceeding in the same way as in section 3 and taking the limit $V_\alpha \to 0$ we get

$$Z^{(0)}_{N,Q} = Z_{\text{EM},Q}(\beta, \hbar)\,\frac{1}{N!}\left(\frac{m}{2\pi\beta\hbar^2}\right)^{3N/2} \sum_P (\mp 1)^{\varepsilon(P)} \int \prod_{i=1}^N d^3\vec{x}_i$$
$$\times \exp\left\{-\beta \sum_{i>j=1}^N \frac{q^2}{4\pi\,|\vec{x}_i(p_i) - \vec{x}_j(p_j)} - \frac{m}{2\beta\hbar^2} \sum_{i=1}^N [\vec{x}_i - \vec{x}_i(Pi)]^2\right\},$$

where the subscript (0) refers to the approximation adopted in (5.1). In the limit $\hbar \to 0$ only the identity permutation contributes to $Z_{N,Q}^{(0)}$, so $Z_{N,Q} \to Z_{N,\mathrm{Cl}}$ as $\hbar \to 0$. The other permutations approach zero faster than the quantum correction for $P = I$ derived in section 4. This agrees with the discussion at the beginning of section 4.

The same result would have been obtained if instead of the canonical ensemble, one uses the grand canonical one, and describe the particles by Schrödinger (Bose) fields with antiperiodic (periodic) boundary conditions.

The advantage of this method is that it keeps memory of the different permutations. On the other hand, and due to the partition of the $\vec{x}$-space into cells and the expansion for $\vec{A}$ in each cell, the interactions among particles mediated by $\vec{A}$ are restricted to take place within the same cell and not among different ones, while in the method of section 4, $\vec{A}$ could propagate the interaction among arbitrary points in the $\vec{x}$-space. This feature, together with the fact that in the limit $\hbar \to 0$ permutations different from the identity tend to zero faster than the quantum corrections for $P = I$, suggest that the method of section 4 is more suitable for evaluating the leading quantum corrections.

6. Conclusions

The important result obtained in section 2 is that in Classical Statistics the dynamical or transverse EM degrees of freedom become decoupled from those of the particles in the partition function. This decoupling between matter and radiation implies that the classical average values for $\vec{A}_{\mathrm{Cl}}$, $\vec{B}_{\mathrm{Cl}}$ and $\vec{E}_{\mathrm{Cl}}$ vanish. Such a decoupling does not take place quantum-mechanically, but it is proved that $Z_{N,Q} \to Z_{N,\mathrm{Cl}}$ as $\hbar \to 0$, which generalizes theorem 15.9 of [11], in which the same property is formulated for an external EM field without dynamical degrees of freedom. Explicit expressions for the leading quantum correction to the classical partition function are given, which constitute the main result of the paper.

Appendix A. Evaluation of Eq. (3.5) and a Bound to Eq. (3.6)

In this appendix we will evaluate the matrix element (3.6) which appears in the partition function (3.5). Eq. (3.6) is a path integral over $\vec{x}_i(\tau)$, $\vec{p}_i(\tau)$ and $q(\vec{k}, \lambda, \tau)$, $\pi(\vec{k}, \lambda, \tau)$ for $i = 1, \ldots, N$ and for all $(\vec{k}, \lambda)$. Making a partition of the time interval $[0, \beta\hbar]$ into $M + 1$ subintervals of length δ,

$$\delta = \frac{\beta\hbar}{M + 1}, \tag{A.1}$$

we have for Eq. (3.6)

$$\begin{aligned}
&\langle \vec{x}_1, \ldots, \vec{x}_N; \{\vec{q}\}| \exp(-\beta H_Q) |\vec{x}_1{}', \ldots, \vec{x}_N{}'; \{\vec{q}\}\rangle \\
&= \int \lim_{M\to\infty} \prod_{l=1}^{M} \prod_{l'=1}^{M+1} \prod_{i=1}^{N} \frac{d^3\vec{x}_{i,l}\, d^3\vec{p}_{i,l'}}{(2\pi\hbar)^3} \prod_{\vec{k},\lambda} \frac{dq_1\, d\pi_{l'}}{2\pi\hbar} \\
&\times \exp\left\{\frac{i}{\hbar} \sum_{l=0}^{M+1} \left[\sum_{i=1}^{N} \vec{p}_{i,l}(\vec{x}_{i,l} - \vec{x}_{i,l-1}) + \int d^3\vec{k}\pi_l(q_l - q_{l-1})\right.\right. \\
&- \delta \sum_{j=1}^{N} \frac{1}{2m} \left[\vec{p}_{j,l} - iq \sum_{\lambda=1,2} \int \frac{d^3\vec{k}}{(2\pi)^{3/2}} c(\vec{k})\, \vec{\varepsilon}(\vec{k}, \lambda)\, q_l \exp\left[\frac{i}{2} \vec{k}(\vec{x}_{j,l} + \vec{x}_{j,l-1})\right]\right]^2 \\
&\left.\left. - \frac{\delta}{2} \int \frac{d^3\vec{k}}{(2\pi)^3} |c(\vec{k})|^2 (|\pi_l|^2 + \omega^2|q_l|^2) - \delta \sum_{i>j=1}^{N} \frac{q^2}{4\pi\, |\vec{x}_{i,l} - \vec{x}_{j,l}|}\right]\right\},
\end{aligned} \tag{A.2}$$

with boundary conditions

$$\vec{x}_{i,M+1} = \vec{x}_i(\beta\hbar) = \vec{x}_i, \quad \vec{x}_{i,0} = \vec{x}_i(0) = \vec{x}_i'; \quad q_{M+1} = q(\beta\hbar) = \tilde{q}, \quad q_0 = q(0) = \tilde{q}.$$

Notice that in Eq. (A.2) we have solved the well-known path integral ambiguity in the coupling of matter to vector potentials by choosing the midpoint $1/2(\vec{x}_{i,l} + \vec{x}_{i,l-1})$ as argument of the EM field in the product $\dot{\vec{x}}_i\vec{A}(\vec{x}_j, t)$ [9, 17]. For a lucid discussion of its relationship to the I to ambiguity see also [17].

Each of the $\vec{p}_{i,l}$ and π_l integrations are gaussian and can be done immediately

$$\langle\vec{x}_1, \ldots, \vec{x}_N; \{\tilde{q}\}|\exp(-\beta H_Q)|\vec{x}_1', \ldots, \vec{x}_N; \{\tilde{q}\}\rangle$$
$$= \int \lim_{M\to\infty} \prod_{l=0}^{M+1}\prod_{i=1}^{N} \frac{d^3\vec{x}_{i,l}}{(m^{-1}2\pi\hbar\delta)^{3/2}} \prod_{\vec{k},\lambda} \frac{dq_l}{(2\pi\hbar\delta)^{1/2}} \exp\left\{\frac{1}{\hbar}\sum_{l=0}^{M+1}\left[-\frac{m}{2\delta}\sum_{i=1}^{N}(\vec{x}_{i,l} - \vec{x}_{i,l-1})^2\right.\right.$$
$$- \frac{\delta}{2}\sum_{\lambda=1,2}\int \frac{d^3\vec{k}}{(2\pi)^3}|c(\vec{k})|^2\left(\left|\frac{q_l - q_{l-1}}{\delta}\right|^2 - \omega^2|q_l|^2\right) - \sum_{i>j=1}^{N}\frac{q^2}{4\pi|\vec{x}_{i,l} - \vec{x}_{j,l}|}$$
$$\left.\left.- q\sum_{i=1}^{N}(\vec{x}_{i,l} - \vec{x}_{i,l-1})\sum_{\lambda=1,2}\int \frac{d^3\vec{k}}{(2\pi)^{3/2}} c(\vec{k})\,\vec{\varepsilon}(\vec{k}, \lambda)\, q_l \exp\left[\frac{i}{2}\vec{k}(\vec{x}_{i,l} + \vec{x}_{i,l-1})\right]\right]\right\}. \tag{A.3}$$

We shall simplify Eq. (A.3) by recalling FEYNMAN's result [15]:

$$\int_{x(0)=x}^{x(U)=x'} \lim_{M\to\infty}\left(\frac{m}{2\pi\hbar\delta}\right)^{1/2}\prod_{l=1}^{M}\frac{dx_l}{(m^{-1}2\pi\hbar\delta)^{1/2}}$$
$$\times \exp\left\{-\frac{1}{\hbar}\sum_{l=0}^{M+1}\delta\left[\frac{m}{2}\left(\frac{x_l - x_{l-1}}{\delta}\right)^2 + \frac{mw^2}{2}x_l^2 + ix_l\gamma_l\right]\right\}$$
$$= \left(\frac{m\omega}{2\pi\hbar\sinh\omega U}\right)\exp\left\{-\frac{1}{4m\omega\hbar}\int_0^U dt\int_0^U dt' e^{-\omega|t-t'|}\gamma(t)\,\gamma(t') - \frac{m\omega}{2\hbar\sinh\omega U}\right.$$
$$\times [(x^2 + x'^2)\cosh\omega U - 2xx' + 2A(xe\omega U - x') + 2B(x'e^{\omega U} - x)$$
$$\left. + (A^2 + B^2)\,e^{\omega U} - 2AB]\right\}, \tag{A.4}$$

with

$$A = \frac{i}{2m\omega}\int_0^U dte^{-\omega t}\gamma(t), \qquad B = \frac{i}{2m\omega}\int_0^U dte^{-\omega(U-t)}\gamma(t). \tag{A.5}$$

Using Eqs. (A.4) and (A.5) to evaluate the integrals over all q_l's in Eq. (A.3), introducing the result into Eq. (3.5) and performing the integration over $\tilde{q}(\vec{k}, \lambda)$, which is also gaussian, we get Eq. (3.7).

Before concluding this section there are some remarks to do about Eq. (3.9).

Remark 1. The exponent in the integrand of (A.3) is the sum of the kinetic energy for the particles (first term), the energy for the free EM field (second term), the Coulomb energy (third term), and the fourth term, which is associated to the interaction between the field and the particles. The last term is purely imaginary, so its exponential has modulus 1. It follows that the propagator (A.3) is bounded by the product of the propagators for N quantum particles interacting via a Coulomb potential and for the free EM field:

$$|\langle \vec{x}_1, \ldots, \vec{x}_N; \{\vec{q}\}| \exp(-\beta H_Q) |\vec{x}_1', \ldots, \vec{x}_N'; \{\vec{q}\}\rangle| \leq \langle \vec{x}_1, \ldots, \vec{x}_N| \exp[-\beta(H^0_{p,Q} + H_C)] |\vec{x}_1, \ldots, \vec{x}_N\rangle \langle \{\vec{q}\}| \exp(-\beta H_{\mathrm{EM},Q}) |\{\vec{q}\}\rangle, \tag{A.6}$$

with $H^0_{p,Q} = \sum_{i=1}^{N} \vec{P}_i^2/2m$, which is the generalization of theorem 15.6 of [11] to dynamical EM fields.

Remark 2. For only one particle ($N = 1$) Eq. (A.6) gives the following bound for the partition function:

$$Z_{N=1,Q} \leq V \left(\frac{m}{2\pi\beta\hbar^2}\right)^{3/2} Z_{\mathrm{EM},Q}(\beta, \hbar). \tag{A.7}$$

If we introduce the system with only one particle in an external potential $\varphi(\vec{x})$ we have

$$Z_{N=1,Q} \leq \left(\frac{m}{2\pi\beta\hbar^2}\right)^{3/2} Z_{\mathrm{EM},Q}(\beta, \hbar) \int d^3\vec{x} \exp[-\beta\varphi(\vec{x})],$$

which is the generalization of theorem 9.2 of [11]. For this case, and calling E_0 the energy of the groun state, we get the bounds

$$\exp(-\beta E_0) \leq \mathrm{Tr}[\exp(-\beta H_Q)] \leq \left(\frac{m}{2\pi\beta\hbar^2}\right)^{3/2} Z_{\mathrm{EM},Q}(\beta, \hbar) \int d^3\vec{x} \exp[-\beta\varphi(x)].$$

Appendix B: Evaluation of (3.7) for $K \to 0$ ($P = I$)

The derivation of Eq. (4.5) proceeds through several steps: 1) Expansion of $G(\vec{y})$ into a power series in $\hbar$, regarding the series formed by all terms of orders $\hbar^n$ with $n \geq 1$ as a perturbation, 2) Expansion of the exponential of such a perturbation into a power series, 3) Study of the order in $\hbar$ of the different terms obtained in the last expansion and selection of the leading quantum correction, 4) Explicit evaluation of these leading quantum corrections.

Step 1. According to what has been said in section 4, in the limit $\hbar \to 0$, the leading contribution comes from the identity permutation $P = I$, for which Eqs. (4.2)–(4.4), together with the fact that $\tau, \tau', |\tau - \tau'| \sim \beta\hbar$, hold. From these it follows that Eq. (3.10) can be written as

$$\begin{aligned} G(\vec{y}) = {} & \frac{m}{2\hbar} \sum_{i=1}^{N} \int_0^{\beta\hbar} d\tau \dot{\vec{\xi}}_i^2(\tau) - \frac{1}{\hbar} \sum_{i>j=1}^{N} \frac{q^2}{4\pi|\vec{x}_i - \vec{x}_j|} \\ & \times \int_0^{\beta\hbar} d\tau\{1 + [\xi_i(\tau) - \xi_j(\tau)]_\alpha D^\alpha_{ij} + [\xi_i(\tau) - \xi_j(\tau)]_\alpha Q^{\alpha\gamma}_{ij} [\xi_i(\tau) - \xi_j(\tau)]_\gamma + \cdots\} \\ & - \frac{q^2}{2\beta\hbar^2} \sum_{i,j=1}^{N} \sum_{\lambda=1,2} \int \frac{d^3\vec{k}}{(2\pi)^3} \frac{|c(\vec{k})|^2}{\omega^2} \int_0^{\beta\hbar} d\tau \int_0^{\beta\hbar} d\tau [\vec{\varepsilon}(\vec{k}, \lambda) \dot{\vec{\xi}}_i(\tau)] \\ & \times e^{i\vec{k}(\vec{x}_i - \vec{x}_j)} \left\{1 + i\vec{k}[\vec{\xi}_i(\tau) - \vec{\xi}_j(\tau')] - \frac{1}{2} [\vec{k}(\vec{\xi}_i(\tau) - \vec{\xi}_j(\tau')]^2 + \cdots\right\} \\ & \times [\vec{\varepsilon}(\vec{k}, \lambda) \dot{\vec{\xi}}_j(\tau')] \left\{2 + \omega^2 \left[\frac{1}{12} (\beta\hbar)^2 + |\tau - \tau'|^2 - \beta\hbar |\tau - \tau'|\right] \right. \\ & \left. - \frac{1}{24} (\omega\beta\hbar)^3 + O((\beta\hbar)^4)\right\}, \end{aligned} \tag{B.1}$$

where the last curly bracket comes from the expansion of $f(\tau - \tau', \beta\hbar)$ into suitable powers and D^{α}_{ij} and Q^{α}_{ij} are the following dipole-moment and quadrupole-tensor contributions:

$$D^{\alpha}_{ij} = \frac{(x_i - x_j)^{\alpha}}{|\vec{x}_i - \vec{x}_j|^2}, \qquad Q^{\alpha\gamma}_{ij} = \frac{3(x_i - x_j)^{\alpha}(x_i - x_j)^{\gamma} - (\vec{x}_i - \vec{x}_j)^2 \delta^{\alpha\gamma}}{|\vec{x}_i - \vec{x}_j|^4},$$
$$(\alpha, \gamma = 1, 2, 3). \tag{B.2}$$

The standard summation conventions for the dummy indices α, β have been used in Eq. (B.1).

Next we study the different terms appearing in Eq. (B.1). The free term (kinetic energy) and the electrostatic energy are of order $\hbar^0$

$$G_0(\vec{\xi}) = G^0(\vec{\xi}) - \beta H_c, \qquad G^0(\vec{\xi}) = -\frac{m}{2\hbar} \sum_{i=1}^{N} \int_0^{\beta\hbar} d\tau \dot{\vec{\xi}}_i^2(\tau), \tag{B.3}$$

with H_c given by Eq. (3.2). There is also a term of order $\hbar^0$ arising from the particle-field interaction, which vanishes by virtue of the boundary condition (4.2), namely

$$-\frac{q^2}{\beta\hbar^2} \sum_{\lambda=1,2} \int \frac{d^3\vec{k}}{(2\pi)^3} \frac{|c(\vec{k})|^2}{\omega^2} \left| \sum_{i=1}^{N} e^{i\vec{k}\vec{x}_i} \int_0^{\beta\hbar} d\tau \vec{\varepsilon}(\vec{k}, \lambda)\, \dot{\vec{\xi}}_i(\tau) \right|^2 = 0.$$

The term linear in $\hbar$ is

$$G_1{}^0(\vec{\xi}) = -\frac{q^2}{2\pi\hbar} \sum_{i>j=1}^{N} \frac{D^{\alpha}_{ij}}{|\vec{x}_i - \vec{x}_j|} \int_0^{\beta\hbar} d\tau \xi_{i,\alpha}(\tau) \tag{B.4}$$

since the other possible contribution

$$G_1{}^F(\vec{\xi}) = -\frac{iq^2}{\beta\hbar^2} \sum_{i,j=1}^{N} \sum_{\lambda=1,2} \int \frac{d^3\vec{k}}{(2\pi)^3} \frac{|c(\vec{k})|^2}{\omega^2}$$
$$\times e^{i\vec{k}(\vec{x}_i - \vec{x}_j)} \int_0^{\beta\hbar} d\tau \int_0^{\beta\hbar} d\tau' [\vec{\varepsilon}(\vec{k}, \lambda)\, \dot{\vec{\xi}}_i(\tau)] \{\vec{k}[\vec{\xi}_i(\tau) - \vec{\xi}_j(\tau')]\} [\vec{\varepsilon}(\vec{k}, \lambda)\, \dot{\vec{\xi}}_j(\tau')]$$

vanishes,

$$G_1{}^F(\vec{\xi}) = 0. \tag{B.5}$$

This can be seen by making in its right-hand side the change $\tau \to \tau'$ (or $i \to j$), or by using the boundary condition (4.2). The term of order $\hbar^2$ is

$$G_2(\vec{\xi}) = G_2{}^Q(\vec{\xi}) + G_2{}^{F1}(\vec{\xi}) + G_2{}^{F2}(\vec{\xi}), \tag{B.6}$$

where

$$G_2{}^Q(\vec{\xi}) = -\frac{q^2}{2\pi\hbar} \sum_{i>j=1}^{N} \frac{Q^{\alpha\gamma}_{ij}}{|\vec{x}_i - \vec{x}_j|} \int_0^{\beta\hbar} d\tau \xi_{i,\alpha}(\tau) [\xi_i(\tau) - \xi_j(\tau)]_{\gamma}, \tag{B.7}$$

$$G_2{}^{F1}(\vec{\xi}) = \frac{q^2}{2\beta\hbar^2} \sum_{i,j=1}^{N} \sum_{\lambda=1,2} \int \frac{d^3\vec{k}}{(2\pi)^3} \frac{|c(\vec{k})|^2}{\omega^2} e^{i\vec{k}(\vec{x}_i - \vec{x}_j)} \int_0^{\beta\hbar} d\tau \int_0^{\beta\hbar} d\tau'$$
$$\times [\vec{\varepsilon}(\vec{k}, \lambda)\, \dot{\vec{\xi}}_i(\tau)] \{\vec{k}[\vec{\xi}_i(\tau) - \vec{\xi}_j(\tau')]\}^2 [\vec{\varepsilon}(\vec{k}, \lambda)\, \dot{\vec{\xi}}_j(\tau')], \tag{B.8}$$

$$G_2^{F2}(\vec{\xi}) = -\frac{q^2}{2\beta\hbar^2}\sum_{i,j=1}^{N}\sum_{\lambda=1,2}\int\frac{d^3\vec{k}}{(2\pi)^3}|c(\vec{k})|^2 e^{i\vec{k}(\vec{x}_i-\vec{x}_j)}\int_0^{\beta\hbar}d\tau\int_0^{\beta\hbar}d\tau'[\vec{\varepsilon}(\vec{k},\lambda)\dot{\vec{\xi}}_i(\tau)]$$
$$\times\left[\frac{1}{12}(\beta\hbar)^2+|\tau-\tau'|^2-\beta\hbar|\tau-\tau'|\right][\vec{\varepsilon}(\vec{k},\lambda)\dot{\vec{\xi}}_j(\tau')]. \tag{B.9}$$

Upon expanding the square and using Eq. (4.2), (B.8) simplifies to

$$G_2^{F1}(\vec{\xi}) = -\frac{q^2}{\beta\hbar^2}\sum_{\lambda=1,2}\int\frac{d^3\vec{k}}{(2\pi)^3}\frac{|c(\vec{k})|^2}{\omega^2}\left|\sum_{i=1}^{N}e^{i\vec{k}\vec{x}_i}\int_0^{\beta\hbar}d\tau[\vec{\varepsilon}(\vec{k},\lambda)\dot{\vec{\xi}}_i(\tau)][\vec{k}\dot{\vec{\xi}}_i(\tau)]\right|^2. \tag{B.10}$$

Regarding $G_2^{F2}(\vec{\xi})$, an integration by parts and the use of Eq. (4.2) lead to

$$G_2^{F2}(\vec{\xi}) = \frac{q^2}{\beta\hbar^2}\sum_{\lambda=1,2}\int\frac{d^3\vec{k}}{(2\pi)^3}|c(\vec{k})|^2\left|\sum_{i=1}^{N}e^{i\vec{k}\vec{x}_i}\int_0^{\beta\hbar}d\tau\vec{\varepsilon}(\vec{k},\lambda)\,\vec{\xi}_i(\tau)\right|^2. \tag{B.11}$$

Summarizing,

$$G = -\beta H_c + G^0 + G_1^D + G_2^Q + G_2^{F1} + G_2^{F2} + O(|\vec{\xi}|^3, \hbar^3), \tag{B.12}$$

where superindices 0, D, Q and F stand for free part dipole, quadrupole and contributions arising from the field. Notice that all the terms included in $O(|\vec{\xi}|^3, \hbar^3)$ are, at least, cubic in $\vec{\xi}$ and $\hbar$.

Step 2. We consider $G_1^D + G_2 + G_2^{F1} + G_2^{F2} + O(|\vec{\xi}|^3, \hbar^3)$ as a perturbation to G_0 and expand the exponential of the former as a power series, that is,

$$Z_{N,Q} = \frac{1}{N!}Z_{\mathrm{EM},Q}(\beta,\hbar)\int\prod_{i=1}^{N}d^3\vec{x}_i\exp(-\beta H_c)\int_0^0\prod_{j=1}^{N}D^3\vec{\xi}_j\exp[G^0(\vec{\xi})]$$
$$\times\sum_{n=0}^{\infty}\frac{1}{n!}[G_1^D(\vec{\xi})+G_2^Q(\vec{\xi})+G_2^{F1}(\vec{\xi})-G_2^{F2}(\vec{\xi})+O(|\vec{\xi}|^3,\hbar^3)]^n. \tag{B.13}$$

Step 3. Using the formula

$$\int_{x(0)=x}^{x(\beta\hbar)=x'}Dx\exp\left\{-\frac{m}{2\hbar}\int_0^{\beta\hbar}\dot{x}^2(t)\,dt\right\} = \left(\frac{m}{2\pi\beta\hbar^2}\right)^{1/2}\exp\left\{-\frac{m}{2\beta\hbar^2}(x'-x'^2)\right\} \tag{B.14}$$

we can integrate the first term of the series of (B.13) and get

$$Z_{N,Q} = \frac{1}{N!}Z_{\mathrm{EM},Q}(\beta,\hbar)\int\prod_{i=1}^{N}d^3\vec{x}_i\exp(-\beta H_c)\left\{\left(\frac{m}{2\pi\beta\hbar^2}\right)^{3N/2}-Z_Q'(\vec{x})\right\}, \tag{B.15}$$

$$Z_Q'(\vec{x}) = \prod_{i=1}^{N}\int_0^0 D^3\xi_i\exp[G(\vec{\xi})]\sum_{n=1}^{\infty}\frac{1}{n!}[G_1^D+G_2^Q+G_2^{F1}+G_2^{F2}+O(|\vec{\xi}|^3,\hbar^3)]^n. \tag{B.16}$$

Notice that the first term of (B.15) is the classical partition function. We have to study the order in $\hbar$ of each of the terms of (B.16). For that purpose we use Eqs. (B.4), (B.7),

(B.11), (B.12), (4.4) and the fact that $\int_0^{\beta\hbar} d\tau \sim \beta\hbar$, and get that the contributions of $(G_1^D)^2$, G_2^0, G_2^{F1} and G_2^{F2} go, respectively, with $\beta^3\hbar^2$, $\beta^2\hbar^2$, $\beta\hbar^2$ and $\beta^2\hbar^2$. Notice that $(G_1^D)^2$ come from $n = 2$ in (B.16). The rest of the terms arising from $n \geqq 2$ and those included in $O(|\vec{\xi}|^3, \hbar^3)$ give contributions going with powers of $\hbar$ greater than 2. Therefore, the first quantum correction is

$$Z'_{Q,2}(\vec{x}) = \prod_{i=1}^{N} \int_0^0 D^3\vec{\xi}_i \exp\left[G^0(\vec{\xi})\right] \left[G_1^D + G_2^Q + G_2^{F1} + G_2^{F2} + \frac{1}{2}(G_1^D)^2\right] \tag{B.17}$$

where subindex 2 recalls the order in $\hbar$.

Step 4. Notice that (B.17), as it stands, is only a formal expression. To evaluate it we write its detailed form, which is the following path integral

$$\begin{aligned} Z^1_{Q,2}(\vec{x}) = & \int \lim_{M\to\infty} \prod_{i=1}^{N} \left(\frac{m}{2\pi\hbar\delta}\right)^{3/2} \prod_{l=1}^{M} \frac{d^3\vec{\xi}_{i,l}}{(m^{-1}\, 2\pi\hbar\delta)^{3/2}} \\ & \times \left[G_1^D + G_2^Q + G_2^{F1} + G_2^{F2} + \frac{1}{2}(G_1^D)^2\right] \\ & \times \exp\left\{-\frac{m}{2\pi\hbar\delta} \sum_{i=1}^{N} \sum_{l=1}^{M+1} (\vec{\xi}_{i,l} - \vec{\xi}_{i,l-1})^2\right\} \end{aligned} \tag{B.18}$$

with

$$\vec{\xi}_{i,M+1} = \vec{\xi}_i(\beta\hbar) = 0, \quad \vec{\xi}_{i,0} = \vec{\xi}_i(0) = 0, \quad \delta = \frac{\beta\hbar}{M+1}, \tag{B.19}$$

and where, now,

$$G_1^D = -\frac{q^2}{2\pi\hbar} \sum_{i>j=1}^{N} \sum_{l=1}^{M+1} \delta \frac{D^\alpha_{ij}}{|\vec{x}_i - \vec{x}_j|} \xi_{i,l,\alpha}, \tag{B.20}$$

$$G_2^Q = -\frac{q^2}{2\pi\hbar} \sum_{i>j=1}^{N} \sum_{l=1}^{M+1} \delta \frac{\xi_{i,l,\alpha} Q^{\alpha\gamma}_{ij}[\xi_{i,l} - \xi_{j,l}]_\gamma}{|\vec{x}_i - \vec{x}_j|}, \tag{B.21}$$

$$G_2^{F1} = -\frac{q^2}{\beta\hbar^2} \sum_{\lambda=1,2} \int \frac{d^3\vec{k}}{(2\pi)^3} \frac{|c(\vec{k})|^2}{\omega^2} I(G_2^{F1}, N, M), \tag{B.22}$$

$$G_2^{F2} = \frac{q^2}{\beta\hbar^2} \sum_{\lambda=1,2} \int \frac{d^3\vec{k}}{(2\pi)^3} |c(\vec{k})|^2\, I(G_2^{F2}, N, M). \tag{B.23}$$

$I(G_2^{F1}, N, M)$ and $I(G_2^{F2}, N, M)$ are given by

$$I(G_2^{F2}, N, M) = \left|\sum_{l=1}^{N} e^{i\vec{k}\vec{x}_i} \sum_{l=1}^{M+1} [\vec{\varepsilon}(\vec{\xi}_{i,l} - \vec{\xi}_{i,l-1})]\, \vec{k}\, \frac{(\vec{\xi}_{i,l} + \vec{\xi}_{i,l-1})}{2}\right|^2, \tag{B.24}$$

$$I(G_2^{F2}, N, M) = \left|\sum_{i=1}^{N} e^{i\vec{k}\vec{x}_i} \sum_{l=1}^{M+1} \delta\vec{\varepsilon}\vec{\xi}_{i,l}\right|^2. \tag{B.25}$$

We recall that for the $\vec{\xi}_i$ coming from the EM field we are using the midpoint rule (see section 3), so for this $\vec{\xi}_i$ we write $(\vec{\xi}_{i,1} + \vec{\xi}_{i,l-1})/2$ instead of $\vec{\xi}_{i,l}$. Notice also that there are $M + 1$ subintervals of length δ in $[0, \beta\hbar]$ and only M non trivial $\vec{\xi}_{i,l}$'s.

To integrate Eq. (B.18) we perform a rotation

$$\begin{pmatrix}\vec{\xi}_{i,1}\\ \vdots\\ \vec{\xi}_{i,M}\end{pmatrix} = \begin{pmatrix} a_{i,11} & \cdots & a_{i,1M}\\ \vdots & & \vdots\\ a_{i,M1} & \cdots & a_{i,MM}\end{pmatrix}\begin{pmatrix}\vec{\chi}_{i,1}\\ \vdots\\ \vec{\chi}_{i,M}\end{pmatrix} \tag{B.26}$$

such that the exponent

$$-\frac{m}{2\pi\hbar\delta}\sum_{i=1}^{N}\sum_{l=1}^{M+1}(\vec{\xi}_{i,l}-\vec{\xi}_{i,l-1})^2 = -\frac{m(M+1)}{2\beta\hbar^2}\sum_{i=1}^{N}(\vec{\xi}_{i,1}\cdots\vec{\xi}_{i,M})\,A_i\begin{pmatrix}\vec{\xi}_{i,1}\\ \vdots\\ \vec{\xi}_{i,M}\end{pmatrix}, \tag{B.27}$$

where A_i is the $M \times M$ matrix

$$A_i = \begin{pmatrix} 2 & -1 & 0 \cdots 0\\ -1 & 2 & -1 \cdots 0\\ 0 & -1 & 2 \cdots 0\\ \vdots & \vdots & \vdots \quad \vdots\\ 0 & 0 & 0 \cdots 2\end{pmatrix},$$

diagonalizes to

$$-\frac{m(M+1)}{2\beta\hbar^2}\sum_{i=1}^{N}\sum_{l=1}^{M}a_{i,l}\vec{\chi}_{i,l}^2.$$

The $a_{i,ml}$ of Eq. (B.26) are the eigenvectors of the matrix A_i with eigenvalues $a_{i,m}$. Since the determinant is invariant under unitary transformations we have that

$$\det A_i = \prod_{l=1}^{M}a_{i,l} = M+1. \tag{B.28}$$

Under the change (B.26) the different G's, Eqs. (B.20)–(B.25), become functions of $\vec{\chi}_{i,l}$ of the same order as the one they had when expressed in terms of $\vec{\xi}_{i,l}$. Then, to perform the integration we only have to use

$$\int_{-\infty}^{+\infty} dx x^{2n} e^{-ax^2} = a^{-(n+(1/2))}\,\Gamma\left(n+\frac{1}{2}\right), \qquad \int_{-\infty}^{+\infty} dx x^{2n+1} e^{-ax^2} = 0. \tag{B.29}$$

In this way we get G_1^D does not contribute to (B.18) since it is antisymmetric in $\vec{\chi}_{i,l}$.

Before integrating (B.18) we are going to study, up to order $\hbar^2$, the influence of the field in the interaction between the particles. For G_1^{F1}, see Eqs. (B.22) and (B.24), it is clear that after some algebra

$$\begin{aligned} I(G_2^{F1}, N, M) = \frac{1}{4}\sum_{i,j=1}^{N}\sum_{l,m=2}^{M} e^{i\vec{k}(\vec{x}_i-\vec{x}_j)}[&(\vec{\varepsilon}\vec{\xi}_{i,l})\,(\vec{k}\xi_{i,l-1})\,(\vec{\varepsilon}\vec{\xi}_{j,m})\,(\vec{k}\vec{\xi}_{j,m-1})\\ &-(\vec{\varepsilon}\vec{\xi}_{i,l})\,(\vec{k}\vec{\xi}_{i,l-1})\,(\vec{\varepsilon}\vec{\xi}_{j,m-1})\,(\vec{k}\vec{\xi}_{j,m})\\ &-(\vec{\varepsilon}\vec{\xi}_{i,l-1})\,(\vec{k}\vec{\xi}_{i,l})\,(\vec{\varepsilon}\vec{\xi}_{j,m})\,(\vec{k}\vec{\xi}_{j,m-1})\\ &+(\vec{\varepsilon}\vec{\xi}_{i,l-1})\,(\vec{k}\vec{\xi}_{i,l})\,(\vec{\varepsilon}\vec{\xi}_{j,m-1})\,(\vec{k}\vec{\xi}_{j,m})]. \end{aligned}$$

Suppose $i \neq j$; then we make the change $\vec{\xi}_{i,l} \to \vec{\xi}_{i,l-1}, \vec{\xi}_{i,l-1} \to \vec{\xi}_{i,l}$ under which the exponential of Eq. (B.18) is invariant, but for which $I(G_2^{F1}, N, M)$ vanishes, so only $i = j$

contributes to the integral. For G_2^{F2} we have that performing the change (B.26) the exponential of Eq. (B.18) becomes symmetric and $I(G_2^{F2} \cdot N, M)$ remains linear in $\vec{\chi}_i$ except for $i = j$, where it is quadratic. So by symmetrical integration we have that only $i = j$ contributes. Thus, we have proved that, up to order $\hbar^2$, the interaction between different particles via the EM field dissappears, but the self-interaction is still present. This leads us to conclude that, in the classical limit, the interaction between different particles is mediated by the Coulomb potential and not by the EM field, in agreement with our results in section 2.

Next, we perform the intergrations in Eq. (B.18). We start with its third term, namely, the one associated to G_2^{F1}, which is the leading quantum correction. By making use of Eqs. (B.27), (B.29) with $n = 1$ and taking into account that in our case $a^{-1} = [2\beta\hbar^2/m(M+1)\, a_{i,l}]$, each integration over all $\vec{\chi}_{i,l}$, together with $a^{-1/2}$, gives a factor

$$\left[\frac{2\pi\beta\hbar^2}{m(M+1)}\right]^{3/2} \left(\frac{1}{a_{i,l}}\right)^{3/2},$$

so the global factor comes out to be

$$\left[\frac{m(M+1)}{2\pi\beta\hbar^2}\right]^{3N/2} \prod_{i=1}^{N} \prod_{l=1}^{M} \left(\frac{1}{a_{i,l}}\right)^{3/2}$$

which, using Eq. (B.28), is equal to

$$\left(\frac{m}{2\pi\beta\hbar^2}\right)^{3N/2}.$$

We still have to deal with the factor a^{-1} of Eq. (B.29), (remember $n = 1$ and we have only used $a^{-1/2}$). For $M = 1$ we have that

$$I(G_2^{F1}, N, M = 1) = 0.$$

For $M = 2$,

$$I(G_2^{F1}, N, M = 2) = \frac{1}{2} \sum_{i=1}^{N} [(\vec{\varepsilon}\vec{\chi}_{i,1})^2 (\vec{k}\vec{\chi}_{i,2})^2 + (\vec{\varepsilon}\vec{\chi}_{i,2})^2 (\vec{k}\vec{\chi}_{i,1})^2],$$

where we have used that only $i = j$ contributes, and then we get a factor

$$\frac{N}{27} \left(\frac{\beta\hbar^2}{3m}\right)^2 \vec{\varepsilon}^2\vec{k}^2.$$

For $M = 3$,

$$I(G_2^{F1}, N, M = 3) = \frac{1}{4} \sum_{i=1}^{N} [(\vec{\varepsilon}\vec{\chi}_{i,1})^2 (\vec{k}\vec{\chi}_{i,2})^2 + (\vec{\varepsilon}\vec{\chi}_{i,2})^2 (\vec{k}\vec{\chi}_{i,1})^2 + (\vec{\varepsilon}\vec{\chi}_{i,1})^2 (\vec{k}\vec{\chi}_{i,3})^2 + (\vec{\varepsilon}\vec{\chi}_{i,3})^2 (\vec{k}\vec{\chi}_{i,1})^2],$$

which yields

$$\frac{N}{8} \left(\frac{\beta\hbar^2}{4m}\right)^2 \vec{\varepsilon}^2\vec{k}^2.$$

In general, for any M we have that the factor coming from a^{-1} is

$$N \left(\frac{\beta\hbar^2}{m}\right)^2 \left(\frac{M-1}{M+1}\right)^3 \vec{\varepsilon}^2\vec{k}^2. \tag{B.30}$$

Since $\vec{\varepsilon}^2(\vec{k}, \lambda) = 1$ and $(M - 1)^3/(M + 1)^3$ tends to 1 for $M \to \infty$ we conclude

$$Z'_{Q,2}(\vec{x}, G_2^{F1}) = -\left(\frac{m}{2\pi\beta\hbar^2}\right)^{3N/2} 2N \frac{\beta\hbar^2}{m^2} \left(\frac{q}{c}\right)^2 \int \frac{d^3\vec{k}}{(2\pi)^3} |c(\vec{k})|^2. \tag{B.31}$$

Analogously we evaluate the corrections going with $\beta^2\hbar^2$, due to G_2^Q and G_2^{F2} (second and fourth terms of Eq. (B.18)). The contribution of G_2^Q goes with the trace of the quadrupole tensor $Q^{ij,\alpha}{}_\alpha$, which vanishes, so

$$Z'_{Q,2}(\vec{x}, G_2^Q) = 0. \tag{B.32}$$

For G_2^{F2} we have

$$Z'_{G,2}(\vec{x}, G_1^{F2}) = \left(\frac{m}{2\pi\beta\hbar^2}\right)^{3N/2} 2N \frac{\beta^2\hbar^2}{m} q^2 \int \frac{d^3\vec{k}}{(2\pi)^3} |c(\vec{k})|^2. \tag{B.33}$$

Finally for $(G_1^D)^2$, which gives rise to the $\beta^3\hbar^2$ correction, we get

$$Z'_{Q,2}(\vec{x}, (G_1^D)^2) = \left(\frac{m}{2\pi\beta\hbar^2}\right)^{3N/2} \left(\frac{q^2}{2\pi}\right)^2 \frac{3}{2} \frac{\beta^3\hbar^2}{m} \sum_{\substack{i>j=1\\ i>j'=1}}^{N} \frac{D^{ij\alpha} D_\alpha{}^{ij'}}{|\vec{x}_i - \vec{x}_j|\, |\vec{x}_i - \vec{x}_{j'}|}. \tag{B.34}$$

For G_2^{F2} and $(G_1^D)^2$ the factor $(M - 1)^3$ of (B.30) is to be replaced by $M^3 + 61/6\, M^2 - 113/3\, M + 27$ and $M^3 - 5M^2 - 9M - 4$, respectively.

Notice that $\vec{\varepsilon}^2\vec{k}^2$ is the only non-trivial invariant under rotations of order four that can be formed out of $\vec{\varepsilon}$ and $\vec{k}$, in agreement with (B.22) and (B.30). Out of ε the only second order non-trivial invariant under rotation is $\vec{\varepsilon}^2$, see Eqs. (B.23) and (B.33). The same happens with $Q^{ij,\alpha}{}_\alpha = 0$, Eqs. (B.21) and (B.32), and with

$$\Delta \equiv \sum_{\substack{i>j=1\\ i>j'=1}}^{N} \frac{D^{ij\alpha} D^{ij'}{}_\alpha}{|\vec{x}_i - \vec{x}_j|\, |\vec{x}_i - \vec{x}_{j'}|},$$

see (B.20) and (B.34).

By plugging Eqs. (B.31)—(B.35) into (B.18) and, in turn, the latter into Eq. (B.15) we get (4.5), which we were looking for.

Appendix C. Proof of Eqs. (4.9)—(4.11)

From Eqs. (3.15), (3.16), (4.2)—(4.4) it follows that

$$C + D \xrightarrow[\hbar\to 0]{} -\frac{q}{\omega} \frac{(c\vec{k})}{(2\pi)^{3/2}} \sum_{i=1}^{N} e^{i\vec{k}\vec{x}_i} \int_0^{\beta\hbar} d\tau [\vec{\varepsilon}\dot{\vec{\xi}}_i(\tau)]\, [i\vec{k}\vec{\xi}_i(\tau)] + O(\hbar^3) \tag{C.1}$$

which tells us that the first quantum correction to $\langle\vec{A}_{\rm Cl}(\vec{x})\rangle$ and $\langle\vec{B}_{\rm Cl}(\vec{x})\rangle$ will come from substituting in Eqs. (3.14) and (3.17) $C + D$ by the right-hand-side of (C.1) and $\exp[G(\vec{y})]$ by $\exp[G^0(\vec{\xi}) - \beta H_c]$:

$$\begin{Bmatrix}\langle\vec{A}_Q(\vec{x})\rangle\\ \langle\vec{B}_Q(\vec{x})\rangle\end{Bmatrix} = -\frac{Z_{{\rm EM},Q}(\beta, \hbar)}{N!\, Z_{N,Q}} \int \prod_{i=1}^{N} d^3\vec{x}_i \exp(-\beta H_c) \prod_{j=1}^{N} \int_0^0 D^3\vec{y}_j \sum_{\lambda=1,2} \int \frac{d^3\vec{k}}{(2\pi)^{3/2}}$$
$$\times \begin{Bmatrix}\vec{\varepsilon}(\vec{k}, \lambda)\\ \vec{k} \wedge \vec{\varepsilon}(\vec{k}, \lambda)\end{Bmatrix} \frac{q}{\omega} c(\vec{k}) \sum_{i=1}^{N} e^{i\vec{k}\vec{x}_i} \int_0^{\beta\hbar} d\tau [\vec{\varepsilon}\dot{\vec{\xi}}_i(\tau)]\, [i\vec{k}\vec{\xi}_i(\tau)]$$
$$\times \exp[G^0(\xi)] + O(\hbar^2)$$

both of which vanish since when performing the path integrals we get a result that goes with $\vec{\varepsilon}(\vec{k}, \lambda)\,\vec{k} = 0$. Therefore, we get Eqs. (4.9) and (4.10).

Regarding the electric field (see Eq. (3.18)), we have that, through the same arguments as those used for the quantum partition function,

$$\langle \vec{E}_Q(\vec{x}) \rangle = \frac{Z_{\mathrm{EM},Q}(\beta, \hbar)}{N!\, Z_{N,Q}} \left(\frac{m}{2\pi\beta\hbar^2}\right)^{3N/2} \int \prod_{i=1}^{N} d^3\vec{x}_i \exp(-\beta H_c)$$

$$\left(\vec{\nabla} \sum_{i=1}^{N} \frac{q}{4\pi\,|\vec{x} - \vec{x}_i|}\right) \left\{1 - \frac{q}{2} N \frac{\beta\hbar^2}{m^2} \mathscr{E} - 3N \frac{\beta^2\hbar^2 c^2}{m} \mathscr{E} + \frac{3\beta^3\hbar^2}{m} \left(\frac{q^2}{2\pi}\right)^2 \Delta + O(\hbar^3)\right\}$$

which leads to Eq. (4.11).

Acknowledgements

The partial financial support given by Comisión Asesora de Investigación Científica y Técnica, Spain, is acknowledged. One of us (F.R.R.) whishes to express his gratitude to the Department of Theoretical Physics of the University of Zaragoza for hospitality.

References

[1] I. Bialynicki-Birula, Ann. Phys. (N.Y.) **67**, 252 (1971); Acta Phys. Austriaca **XVIII**, 111 (1977).
[2] F. Ruiz Ruiz and R. F. Alvarez-Estrada, Ann. l'Inst. Henri Poincarè **41**, 143 (1984).
[3] G. Dente, Phys. Rev. **D 12** (1733) (1975).
[4] H. Stapp, Phys. Rev. **D 28**, 1386 (1983).
[5] M. B. Halpern and W. Siegel, Phys. Rev. **D 16**, 2486 (1977).
[6] F. Rohrlich, "Fundamental Physical Problems of Quantum Electrodynamics", in Foundations of Radiation Theory and Quantum Electrodynamics, edited by A. O. Barut, Plenum Press, New York (1980).
[7] C. Bernard, Phys. Rev. **D 9**, 3312 (1974).
[8] L. Dolan and R. Jackiw, Phys. Rev. **D 9**, 3320 (1974); I. Ojima, Ann. Phys. (N.Y.) **137**, 1 (1981); A. J. Nieni and E. W. Semenoff, Ann. Phys. (N.Y.) **152**, 105 (1984); H. Matsumoto, I. Ojima and H. Umezawa, Ann. Phys. (N.Y.) **152**, 348 (1984); H. Matsumoto, Fortschr. Phys. **25**, 1 (1977); M. B. Kislinger and P. D. Morley, Phys. Rev. **D 13**, 2765 (1976); P. D. Morley, Phys. Rev. **D 17**, 518 (1978);
[9] R. P. Feynman, Rev. Mod. Phys. **20**, 367 (1948).
[10] R. P. Feynman, and A. R. Hibbs, "Quantum Mechanics and Path Integrals", McGraw-Hill, New York (1965).
[11] B. Simon, "Functional Integration and Quantum Physics", Academic Press, New York (1979), and references therein.
[12] W. Thirring: "Quantum Mechanics of large systems", Springer, New York (1983).
[13] R. Balescu, "Equilibrium and non-equilibrium Statistical Mechanics", John Wiley & Sons, New York (1975).
[14] S. Ichimaru, "Basic principles of plasma physics", W. A. Benjamin Inc., Reading, Mass. (1973).
[15] R. P. Feynman, "Statistical Mechanics", W. A. Benjamin Inc., Reading Mass. (1972).
[16] K. Huang, "Statistical Mechanics", John Wiley, New York (1963).
[17] L. S. Schulman, "Techniques and Applications of Path Integration", John Wiley & Sons, New York (1981).

[18] L. D. LANDAU and E. LIFSCHITZ, "Statistical Physics", 3rd edition, part 1, Pergamon Press, London (1980).
[19] W. WIGNER, Phys. Rev. **40**, 479 (1932); J. G. KIRWOOD, Phys. **44**, 31 (1933); M. GOLDBERGER and E. N. ADAMS, J. Chem. Phys. **20**, 240 (1952).
[20] E. H. LIEB and W. THIRRING, Phys. Rev. Lett. **35**, 687 (1975); J. M. COMBES, R. SCHRADER and R. SEILER, Ann. Phys. **111**, 1 (1978); K. HEPP and E. H. LIEB; Phys. Rev. **A 8**, 2517 (1973).
[21] S. S. SCHWEBER, "An Introduction to Relativistic Quantum Field Theory", Harper and Row, New York (1964).
[22] In preparation.

Fortschr. Phys. 34 (1986) 11, 775–790

Perturbation Theory, Level-Splitting and Large-Order Behaviour

H. J. W. Müller-Kirsten

Department of Physics, University of Kaiserslautern,
6750 Kaiserslautern, Federal Republic of Germany

Abstract

The level-splitting of asymptotically degenerate eigenvalues, their discontinuity across a cut and the large-order behaviour of the perturbation expansion are usually difficult to calculate. In the following we review some methods of calculation with particular emphasis on the periodic potential in order to exhibit the relationship between all three quantities.

1. Introduction

It is a feature of a physicist's approach to a problem that he employs a method of approximation which is such that the first term of the corresponding expansion yields a rough value of the desired answer. This approach leads automatically to asymptotic series. The characteristic difference between asymptotic and convergent series is illustrated by the following example taken from ref. [1]. The function $\mathrm{Ei}_p(z)$ possesses the convergent expansion

$$\mathrm{Ei}_p(z) \equiv \int_1^\infty u^{-p} e^{-uz} du$$
$$= \sum_{r=0}^{\infty} \frac{(-z)^r}{r!(p-r-1)} + \frac{\pi z^{p-1}}{(p-1)!\sin \pi p} \qquad (p \neq \text{integer}) \tag{1}$$

and the asymptotic expansion

$$\mathrm{Ei}_p(z) \simeq \frac{e^{-z}}{z} \sum_{r=0}^{\infty} \frac{(p+r-1)!}{(p-1)!\,(-z)^r}. \tag{2}$$

The behaviour of the rth term (r large) of the convergent expansion is seen to be of the form of a power divided by a factorial; in the case of the asymptotic expansion it is the opposite: a factorial in r divided by a power. The difference in the behaviour of the absolute magnitudes of successive terms of the expansions is illustrated in Fig. 1 [1].

The formal perturbation expansion of a quantity $E(g^2)$, i.e.

$$E(g^2) = \sum_{n=0}^{\infty} E_n g^{2n}$$

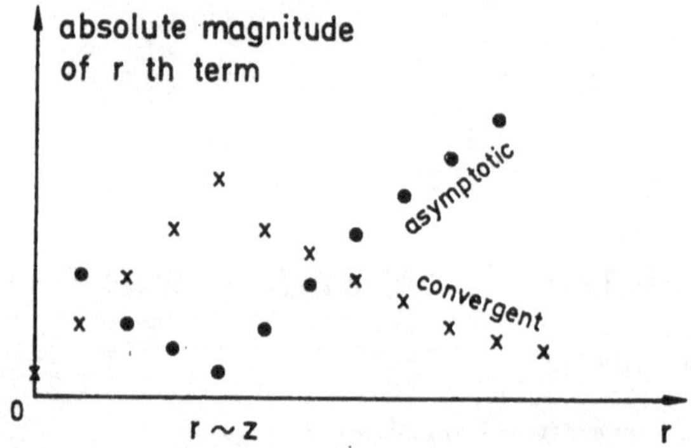

Fig. 1. Typical behaviour of absolute magnitude of successive terms of asymptotic and convergent series

is an asymptotic series if for any N

$$\lim_{g^2\to 0} \frac{1}{g^{2N}} \left(E(g^2) - \sum_{n=0}^{N} E_n g^{2n} \right) = 0$$

or, equivalently for $g = 1/h$,

$$\lim_{h\to\infty} h^{2N} \left(E(h^2) - \sum_{n=0}^{N} \frac{E_n}{h^{2n}} \right) = 0 . \tag{3}$$

Thus for any given value of g^2, $E(g^2)$ is given only approximately by $\sum_{n=0}^{N} E_n g^{2n}$, even if each E_n is known. The traditional method of using an asymptotic expansion is to truncate the series at the least term; this term represents the size of the error (for an authoritative discussion of these aspects we refer to DINGLE [2]).

A considerable amount of work has recently been devoted to the question as to whether it is possible to reconstruct a function exactly from its asymptotic expansion. In some cases the answer is affirmative, and the expansion is said to be Borel summable. If $E(g^2)$ can be written as the Laplace transform

$$E(g^2) = \int_0^\infty dz e^{-z/g^2} F(z) . \tag{4}$$

$F(z)$ is called the Borel transform of E and the integral $E(g^2)$ is called the Borel sum. The question arises: what is the information about $F(z)$ which can be obtained from the asymptotic expansion of E, i.e.

$$E(g^2) = \sum_{n=0}^{\infty} E_n g^{2n} . \tag{5}$$

Asymptotic expansions originate from integrating a series (i.e. that of $F(z)$) over a domain that is larger than its region of convergence (see e.g. ref. [2]). Thus if

$$F(z) = a_{-1}\delta(z) + \sum_{n=0}^{\infty} a_n z^n$$

then with

$$\Gamma(z) = \int_0^\infty e^{-t} t^{z-1}\, dt$$

$$a_{-1} = E_0, \qquad a_0 = E_1/0!, \ldots, a_n = E_{n+1}/n!.$$

Now if

$$E_{n+1} \simeq n!\,(-1)^n \qquad (\text{and } E_0 = 0)$$

$$F(z) \simeq \sum_{n=0}^{\infty} (-z)^n = \frac{1}{1+z}. \tag{6}$$

This power series of F converges inside the unit circle about the origin at $z = 0$; $F = 1/(1+z)$ is an analytic function which can be continued beyond this circle to the entire positive z-axis as required by (4). In this ideal case the integral of (4) can be evaluated and $E(g^2)$ thus recovered from the asymptotic expansion by Borel summation. In general this is not so easy. In fact, many asymptotic series do not even alternate in sign (as in (6)). In any case: it is important to know the large-order behaviour of the asymptotic series so that the question of its (exact or only approximate) Borel summability can be decided.

We observe that a function $E(h)$ with asymptotic series

$$-\sum_{r=1}^{\infty} A_r/h^r \tag{7}$$

has an essential singularity at $h = 0$. We can therefore write down a dispersion relation representation choosing the cut from 0 to $+\infty$:

$$E(h) = \frac{1}{\pi} \int_0^\infty \frac{\operatorname{Im} E(h')}{h' - h}\, dh' \tag{8}$$

except for possible subtractions. Comparing (7) and (8) we obtain

$$A_r = \frac{1}{\pi} \int_0^\infty h'^{r-1} \operatorname{Im} E(h')\, dh'. \tag{9}$$

The question is therefore; how can we determine $\operatorname{Im} E(h)$, or can we determine A_r independently?

Now, $\operatorname{Im} E \neq 0$ in a problem where probability can leak away, i.e. if the potential or equivalently a boundary condition is complex such that the Hamiltonian is nonself-adjoint. This is the case if the real potential has a hump of some sort, e.g. if

$$V(r) = -\frac{1}{r} + gr^N, \qquad N = 1, 2, \ldots$$

as shown in Fig. 2 and $g < 0$ (this example has been considered in detail in ref. [3]). Im E can be calculated by the WKB procedure as in ref. [4], where it is shown that with $g = 1/h$ and

$$E = \frac{p^2}{2m} + V(r)$$

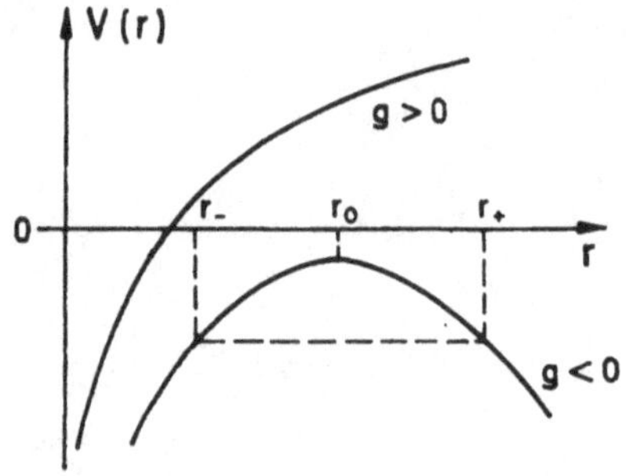

Fig. 2. The potential $V(r) = -1/r + gr^N$, $N = 1, 2, \ldots$ for $g \gtrless 0$

one finds

$$\begin{aligned} E_n &= \varepsilon_0^{(n)} - \frac{i}{2}\,\gamma(g) \\ \gamma(g) &\simeq \gamma_0 \exp\left\{-2\int_{r_-}^{r_+} |p|\,dr\right\} \\ \gamma_0 &= \gamma_0(n) \simeq \omega_{\text{classical}}/2\pi \\ \omega_{\text{classical}} &= 2\pi/T_{\text{Kepler}} = 1/n^3. \end{aligned} \tag{10}$$

Here γ_0 is the frequency of particle bounces against the potential barrier. Inserting $\operatorname{Im} E = -1/2\gamma$ in A_r and evaluating the integral (e.g. by the method of steepest descent) for the Yukawa potential

$$V(r) = -ge^{-r}/r$$

the authors of ref. [4] obtain

$$E = \sum_{r=0}^{\infty} E_r g^r, \qquad E_r \simeq (-1)^{r+1}\, r!\left(\frac{n}{2\ln r}\right)^r.$$

The logarithmic factor in E_r is a novel feature of this case.

In the following we consider a different method, exploiting the asymptotic perturbation expansion. In particular we consider a periodic potential and calculate in turn the level-splitting, the imaginary part of the eigenvalues (for nonselfadjoint boundary conditions) and the large-order behaviour of the perturbation expansion. Our emphasis is not so much on the individual cases which have been considered earlier, but on their interrelationship.

2. Perturbation Expansions: Oscillator-Type and WKB-Type

We distinguish two types of expansions: oscillator-type expansions which are valid in the immediate neighbourhood of an extremum, and WKB-type expansions which are valid beyond this neighbourhood (the matching of the latter to WKB solutions has been demonstrated in ref. [5]). The ideal example to consider is the periodic potential. Perturbation solutions of the equation

$$\psi'' + \{\lambda - 2h^2 \cos 2x\}\,\psi = 0 \qquad \left(-\frac{\pi}{2} \leq x \leq \frac{\pi}{2}\right) \tag{11}$$

have an exceptional property: the coefficients of the perturbation expansions are the same for both types of expansions [6]. This means the following. We set

$$\psi = \varphi(x)\, e^{2h \sin x}, \qquad \lambda + 2h^2 = 2hq + \frac{\Delta}{8}. \tag{12}$$

Here q is only approximately $q_0 = 2n + 1$, $n = 0, 1, \dots$ depending on boundary conditions. We shall be interested in the deviation of q from an exact integer, i.e. $q - q_0$. Then [6]

$$\varphi = \varphi_q + \sum_{i=1}^{\infty} \frac{1}{(2^7 h)^i} \sum_{\substack{j=-i \\ j \neq 0}}^{i} P_i(q, j)\, \varphi_{q+4j} \tag{13}$$

together with

$$-2\Delta = \sum_{i=1}^{\infty} \frac{M_i}{(2^7 h)^{i-1}}, \qquad M_i = M_i(P_j).$$

The coefficients $P_i(q, j)$ are the perturbation coefficients mentioned above.
For the oscillator-type case about $x \sim \pi/2$ we have

$$\varphi_q(x) = \frac{H_{1/2(q-1)}(z)}{2^{1/4(q-1)} \left[\frac{1}{4}(q-1)\right]!} \equiv B_q(z(x))$$

$$z(x) = 4h^{1/2} \cos\left(\frac{1}{4}\pi + \frac{1}{2}x\right)$$

valid for $|z| \ll h^{1/2}$, $|\cos(1/4\pi + 1/2x)| \ll 1$,

$$\text{i.e. about } x \sim \frac{\pi}{2}. \tag{14}$$

For the oscillator-type case about $x \sim -\pi/2$ we have

$$\varphi_q(x) = 2^{1/4(q+1)} \left[\frac{1}{4}(q-3)\right]!\, H^*_{-1/2(q+1)}(z) \equiv C_q(z(x))$$

$$\left(H^*(z) = (-i)^m H_m(iz)\right) \tag{15}$$

$$z(x) = 4h^{1/2} \cos\left(\frac{1}{4}\pi - \frac{1}{2}x\right)$$

valid for $|z| \ll h^{1/2}$, $|\cos(1/4\pi - 1/2x)| \ll 1$, i.e. about $x \sim -\pi/2$.
For the WKB-type case away from $x \sim \pi/2$ we have

$$\varphi_q(x) = \frac{\cos^{1/2(q-1)}\left(\frac{1}{4}\pi + \frac{1}{2}x\right)}{\sin^{1/2(q+1)}\left(\frac{1}{4}\pi + \frac{1}{2}x\right)} = \frac{2^{1/2}}{(\cos x)^{1/2}} \exp\left(-\frac{q}{2}\int \frac{dx}{\cos x}\right) \equiv A_q(x) \tag{16}$$

valid for

$$\left|\cos\left(\frac{1}{4}\pi + \frac{1}{2}x\right)\right| \gg h^{-1/2}, \text{ i.e. away from } x \sim \frac{\pi}{2}.$$

The various domains are shown in Fig. 3. The coefficients P_i are the same in each case.

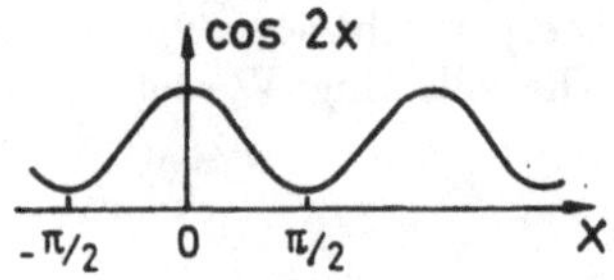

Fig. 3. The periodic potential

In each of the above cases a second linearly independent solution $\bar{\psi}$ is obtained by the replacement $x \to -x$. Thus we have pairs

$$\left.\begin{aligned} \psi_A(x) &= e^{2h\sin x}\left[A_q(x) + o\left(\frac{1}{h}\right)\right] \\ \bar{\psi}_A(x) &= e^{-2h\sin x}\left[\bar{A}_q(x) + o\left(\frac{1}{h}\right)\right], \quad \bar{A}_q(x) = A_q(-x) \end{aligned}\right\} \tag{17}$$

$$\left.\begin{aligned} \psi_B(x) &= e^{2h\sin x}\left[B_q(x) + o\left(\frac{1}{h}\right)\right] \\ \bar{\psi}_B(x) &= e^{-2h\sin x}\left[\bar{B}_q(x) + o\left(\frac{1}{h}\right)\right], \quad \bar{B}_q(x) = B_q(z\,(-x)) \end{aligned}\right\} \tag{18}$$

and similarly a pair $\psi_c(x)$, $\bar{\psi}_c(x)$. We then have solutions with the domains indicated in Fig. 4.

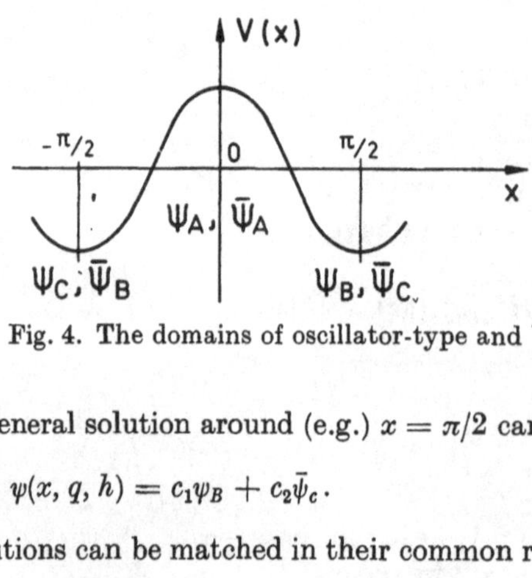

Fig. 4. The domains of oscillator-type and WKB-type solutions

The general solution around (e.g.) $x = \pi/2$ can be written

$$\psi(x, q, h) = c_1\psi_B + c_2\bar{\psi}_c \,. \tag{19}$$

The solutions can be matched in their common region of validity or overlap. Thus [6]

$$\psi_B = \alpha\psi_A\,, \qquad \bar{\psi}_c = \bar{\alpha}\bar{\psi}_A \tag{20}$$

where [6]

$$\left.\begin{aligned} \alpha &= \frac{(8h)^{1/4(q-1)}}{\left[\frac{1}{4}\,(q-1)\right]!}\left(1 + o\left(\frac{1}{h}\right)\right) \\ \bar{\alpha} &= \frac{\left[\frac{1}{4}\,(q-3)\right]!}{(8h)^{1/4(q+1)}}\left(1 + o\left(\frac{1}{h}\right)\right). \end{aligned}\right\} \tag{21}$$

3. Even and Odd Solutions and the Level-Splitting

We obtain real even and odd solutions with the combinations

$$\psi_\pm = A(x)\, e^{2h \sin x} \pm \bar{A}(x)\, e^{-2h \sin x} \tag{22}$$

(ignoring an overall constant). The continuation of these solutions to the domain around the extremum at $x_0 = \pi/2$, $z = 0$, is given by

$$\psi_\pm = \frac{B[z(x)]}{\alpha}\, e^{2h \sin x} \pm \frac{\bar{C}[z(x)]}{\bar{\alpha}}\, e^{-2h \sin x}. \tag{23}$$

We now impose periodicity boundary conditions on $\psi_\pm$ (these boundary conditions are selfadjoint and hence lead to real eigenvalues); e.g. such that the Wronskian of ψ_-, ψ_+ is 1:

$$(\psi_+)_{x_0} = +1, \quad (\psi_-)_{x_0} = 0, \quad \left(\frac{\partial \psi_+}{\partial x}\right)_{x_0} = 0, \quad \left(\frac{\partial \psi_-}{\partial x}\right)_{x_0} = -1. \tag{24}$$

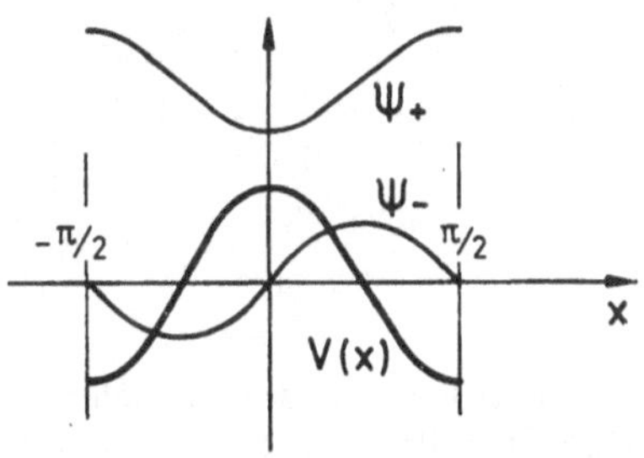

Fig. 5. Typical shape of even and odd solutions in $[-\pi/2, \pi/2]$

The shape of typical solutions of this type is shown in Fig. 5. Evaluating (24) one obtains $\tan \pi/4(q-1) .= \cdots$; expanding the tangent about q_0 leads to [6]

$$q - q_0 = \pm 2 \left(\frac{2}{\pi}\right)^{1/2} \frac{(16h)^{1/2q_0}\, e^{-4h}}{\left[\frac{1}{4}(q_0-1)\right]!} \left\{1 + o\left(\frac{1}{h}\right)\right\} \tag{25}$$

the upper sign referring to ψ_-, the lower to ψ_+. The splitting of the eigenvalues follows immediately with

$$\begin{aligned} \lambda(q) &\simeq \lambda(q_0) + (q - q_0) \left(\frac{\partial \lambda}{\partial q}\right)_{q_0} \\ &= \lambda(q_0) \pm \frac{(16h)^{1/2q_0+1}\, e^{-4h}}{(8\pi)^{1/2} \left[\frac{1}{2}(q_0-1)\right]!} \left\{1 + o\left(\frac{1}{h}\right)\right\}. \end{aligned} \tag{26}$$

This shows the expected band structure.

As mentioned before, the beauty of the perturbation solutions for the periodic potential is that the coefficients $P_i(q, j)$ are the same for oscillator-type as well as WKB-type solutions. This is exceptional. In many other cases, e.g. for the double-well potential, it is more convenient to use parabolic cylinder functions $D_n(z)$ instead of Hermite H_n. Thus,

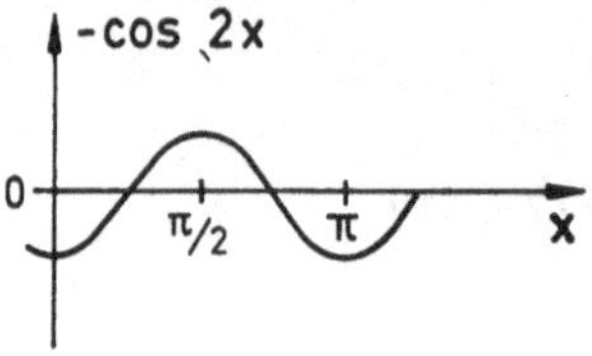

Fig. 6. The periodic potential shifted by $\pi/2$

replacing x of (11) by $x \pm \pi/2$ we obtain

$$\psi'' + \{\lambda + 2h^2 \cos 2x\} \psi = 0 \tag{27}$$

with the potential $-\cos 2x$ shown in Fig. 6. Around $x \sim 0$:

$$\psi'' + \{\lambda + 2h^2 - 4h^2x^2 + \cdots\} \varphi = 0 .$$

Setting

$$\omega = 2h^{1/2}x, \quad \lambda + 2h^2 = 2hq + \frac{\Delta}{8} \tag{28}$$

we have

$$\frac{d^2\psi(\omega)}{d\omega^2} + \left\{\frac{1}{2} q - \frac{\omega^2}{4}\right\} \psi(\omega) \simeq 0 . \tag{29}$$

The solutions of this equation are

$$D_{1/2(q-1)}(\pm\omega), \quad D_{-1/2(q+1)}(\pm i\omega) .$$

Using expansions in terms of these functions in the case of the double-well potential and correspondingly WKB-type expansions, the level-splitting can be calculated and is found to be very similar to (26) (see refs. [7], [8]).

4. Imaginary Part of Eigenvalue for the Periodic Potential

In order to obtain the discontinuity of λ we have to go to nonselfadjoint boundary conditions so that λ is no longer real (continuation of λ into the complex plane). However, we want these to be such that the same solutions are retained. One way to do this is to continue the even (odd) solution defined in the domain of one trough of the potential to a finite complex solution in the domain of a neighbouring trough. The latter can be expressed as the boundary condition (cf. ref. [9])

$$\psi(\pm i\omega) \to 0 \quad \text{for} \quad \left.\begin{array}{l} \omega = 2h^{1/2}x \\ x \to \pi \pm i\infty . \end{array}\right\} \tag{30}$$

We consider first the ground state, for which $q \simeq q_0 = 1$. Replacing x in ψ_A, $\overline{\psi}_A$ of (17) by $x + \pi/2$ (this being the shift described earlier) we obtain

$$\psi_A \simeq \frac{e^{2h \cos x}}{\cos \frac{1}{2} x}, \quad \overline{\psi}_A \simeq \frac{e^{-2h \cos x}}{-\sin \frac{1}{2} x} \tag{31}$$

as the WKB-type solutions. The corresponding oscillator-type solutions are

$$\psi\colon D_{1/2(q-1)}(\pm\omega), \quad D_{-1/2(q+1)}(\pm i\omega) . \tag{32}$$

We set

$$\lambda = E_1 + E_2 = -2h^2 + 2hq + \frac{\Delta}{8}$$

$$E_1 = -2h^2 + 2hq_0 + \frac{\Delta}{8} \tag{33}$$

$$\frac{1}{2}(q - q_0) = \frac{E_2}{4h}.$$

For the ground state with $q_0 = 1$ the even (symmetric) solution $\psi \equiv \psi_s$ can be defined by (cf. ref. [9])

$$\psi_s = \frac{1}{2}\left[D_{E_2/4h}(2h^{1/2}x) + D_{E_2/4h}(-2h^{1/2}x)\right]$$

$$\simeq D_0(2h^{1/2}x) + E_2 \frac{\partial}{\partial E_2} \frac{1}{2}\left[D_{E_2/4h}(2h^{1/2}x) + D_{E_2/4h}(-2h^{1/2}x)\right]$$

about $E_2 = 0$. Near $x = 0$

$$\psi_S \simeq D_0(2h^{1/2}x) + \frac{1}{2} E_2 \frac{\partial}{\partial E_2}\left[e^{iE_2\pi/4h} \frac{(2h^{1/2}x)^{-(1+(E_2/4h))} e^{hx^2}(2\pi)^{1/2}(-1)^{-E_2/4h}}{\Gamma(-E_2/4h)}\right]$$

$$\simeq D_0(2h^{1/2}x) - \frac{E_2(2\pi)^{1/2} e^{hx^2}}{16h^{3/2}x}. \tag{34}$$

We now match this to a linear combination of $q = 1$ WKB-type solutions, i.e. (31), i.e.

$$\psi_{\text{WKB}} \simeq c_1 \frac{e^{2h\cos x}}{\cos\frac{1}{2}x} + c_2 \frac{e^{-2h\cos x}}{-\sin\frac{x}{2}}. \tag{35}$$

Now, near $x = 0$:

$$D_0(2h^{1/2}x) = e^{-hx^2} \sim e^{-2h}\left(\frac{e^{2h\cos x}}{\cos\frac{1}{2}x}\right).$$

Choosing $c_1 = e^{-2h}$ we obtain

$$\psi_{\text{WKB}} \simeq D_0(2h^{1/2}x) + \frac{c_2 e^{-2h\cos x}}{-\sin\frac{x}{2}} \simeq D_0(2h^{1/2}x) + \frac{c_2 e^{-2h} e^{hx^2}}{-\frac{x}{2}} \tag{36}$$

near $x = 0$. Next we determine c_2 by considering ψ_{WKB} near $x = \pi$. To this end we set $x' = x - \pi$. Then

$$\psi_{\text{WKB}} \simeq -\frac{c_1 e^{-2h} e^{hx'^2}}{\frac{1}{2}x'} - c_2 e^{2h} e^{-hx'^2}. \tag{37}$$

Now near $x = \pi$ the boundary condition (30) demands [9]

$$\psi = AD_{-1}(\pm i2h^{1/2}x') \to 0 \quad \text{for} \quad x' \to \pm i\infty \tag{38}$$

(recall that $D_\nu(\omega) \simeq \omega^\nu e^{-\omega^2/4}$).

The asymptotic behaviour of $D_{-1}(\pm i\omega)$ is obtained from tables. Then

$$\psi \simeq A \left(\frac{\pi}{2}\right)^{1/2} e^{-hx'^2} \mp Ai \frac{e^{hx'^2}}{2h^{1/2}x'}. \tag{39}$$

Equating (37) and (39) we obtain

$$\left.\begin{aligned} A &= \pm \frac{4h^{1/2}}{i} e^{-4h} \\ c_2 &= \mp \frac{1}{i}\, 2(2h\pi)^{1/2}\, e^{-6h}. \end{aligned}\right\} \tag{40}$$

Hence (36) can be written

$$\psi_{\text{WKB}} = D_0(2h^{1/2}x) \pm \frac{4}{i} (2h\pi)^{1/2} e^{-8h} \frac{e^{hx^2}}{x}. \tag{41}$$

Comparing this with ψ_s we obtain

$$E_2 \equiv i \operatorname{Im} \lambda = \pm i h^2 2^6 e^{-8h}. \tag{42}$$

We can repeat these calculations for the general case (considering both even and odd wave functions around $x = 0$) and obtain

$$E_2 \simeq \pm i \frac{(16h)^{q_0+1}\, e^{-8h}}{4\left\{\left[\frac{1}{2}(q_0 - 1)\right]!\right\}^2}. \tag{43}$$

Comparing this expression with the level-splitting determined by (25) we have (using $E = \lambda = E_1 + E_2$, $E_1 = -2h^3 + 2hq + \cdots$).

$$\operatorname{Im} E(h) \simeq \left(\frac{\partial E}{\partial q}\right)_{q_0} \pi \left(\frac{q - q_0}{2}\right)^2. \tag{44}$$

For the double-well potential a similar formula is cited in ref. [10].

5. Large-Order Behaviour of Perturbation Expansion for the Periodic Potential

The perturbation expansion of the eigenvalues can be written

$$\lambda = -2h^2 + 2hq + \frac{\Delta}{8}, \qquad \Delta = -\frac{1}{2} \sum_{r=0}^{\infty} \frac{M_{r+1}}{(2^7h)^r}$$

or

$$\lambda + 2h^2 - 2hq - \frac{M_1}{2^4} = -\frac{1}{2^4} \sum_{r=1}^{\infty} \frac{M_{r+1}}{(2^7h)^r} = -\sum_{r=1}^{\infty} \frac{A_r}{h^r}. \tag{45}$$

We can calculate A_r for large r from

$$A_r = \frac{1}{\pi} \int_0^{\infty} h'^{r-1} \operatorname{Im} E(h')\, dh'$$

with Im $E(h)$ from (43), i.e.

$$\operatorname{Im} E(h) = \frac{(16h)^{q_0+1}\, e^{-8h}}{4\left\{\left[\frac{1}{2}(q_0-1)\right]!\right\}^2}. \tag{46}$$

Then

$$A_r \simeq -\frac{2^{2n}(r+2n+1)!}{\pi \cdot 2^{3r}(n!)^2} \tag{47}$$

where

$$n = \frac{1}{2}(q_0 - 1).$$

We can check this result. In ref. [11] the large-order behaviour has been calculated by solving the recurrence relation of the perturbation coefficients P_i as a difference equation. There it was found that for large r

$$r^{\text{th}} \text{ term} \simeq -\frac{r!\, r^{q-1}}{\left\{\left[\frac{1}{4}(q-1)\right]!\left[\frac{1}{4}(q-3)\right]!\right\}^2 (8h)^{r-1}}. \tag{48}$$

Using

$$\frac{r!}{(r+m)!} \overset{r\,\text{large}}{\simeq} \frac{1}{r^m}\left\{1 - o\left(\frac{1}{r}\right)\right\} \tag{49}$$

and

$$\Gamma(z)\,\Gamma\left(z+\frac{1}{2}\right) = \frac{\Gamma(2z)\,\pi^{1/2}}{2^{2z-1}}$$

this becomes

$$r^{\text{th}} \text{ term} \simeq -\frac{(r+2n)!\, z^{2n}}{\pi(8h)^{r-1}(n!)^2} = \frac{A_{r-1}}{h^{r-1}}. \tag{50}$$

We observe that successive terms do not alternate in sign. Thus exact Borel summation is not possible.

An alternative derivation of the large-order behaviour of the eigenvalue expansion for the periodic potential has been found by DINGLE [1]. We sketch this here slightly modified. We consider again the periodic potential with eigenfunction expansion

$$\psi = \psi_q + \sum_{i=1}^{\infty} \frac{1}{(2^7h)^i} \sum_{\substack{j=-i \\ j\neq 0}}^{i} P_i(q,j)\, \psi_{q+4j}. \tag{51}$$

Inserting this into the wave equation we obtain the recurrence relation (cf. ref. [12])

$$\begin{aligned} jP_i(q,j) = {} & P_{i-1}(q,j-1)\,(q+4j-1)\,(q+4j-3) \\ & + P_{i-1}(q,j)\cdot 2[(q+4j)^2 + 1 + \Delta] \\ & + P_{i-1}(q,j+1)\,(q+4j+1)\,(q+4j+3). \end{aligned} \tag{52}$$

It can be shown [6] that the coefficients M_r of the eigenvalue expansion (45) are given by

$$M_{2r} = 2\sum_{j=1}^{r} j[P_r(q,j)]^2 \simeq M_{2r+1} \quad \text{for large } r. \tag{53}$$

It is convenient to define

$$p_i(\alpha) \equiv \frac{\alpha}{(2^7 h)^i} P_i(q, j), \qquad \alpha = j + \frac{1}{2} q \tag{54}$$

and to consider large values of j for large values of i. Taking dominant terms, e.g. $(q + 4j)^2 \simeq 4j(4j + 2q)$, (52) becomes

$$\frac{8h}{\alpha} p_i(\alpha) = p_{i-1}(\alpha - 1) + 2p_{i-1}(\alpha) + p_{i-1}(\alpha + 1). \tag{55}$$

We now consider

$$p(\alpha) = \sum_i p_i(\alpha) \quad \text{with} \quad p_i(\alpha) \propto \frac{1}{h^i} \tag{56}$$

and

$$p(\alpha - 1) + p(\alpha + 1) = \left(-2 + \frac{8h}{\alpha}\right) p(\alpha). \tag{57}$$

Taking terms of $o(h^{-i+1})$ of (57) we regain (55). Now, the recurrence relation of Whittaker functions W is [13]

$$(2\alpha - z)\, W_{\alpha,1/2}(z) + W_{\alpha+1,1/2}(z) = (1 - \alpha)\, \alpha W_{\alpha-1,1/2}(z).$$

Setting

$$V_{\alpha,1/2}(z) \equiv \frac{W_{\alpha,1/2}(z)}{(\alpha - 1)!} \tag{58}$$

we have

$$V_{\alpha-1,1/2}(z) + V_{\alpha+1,1/2}(z) = \left(-2 + \frac{z}{\alpha}\right) V_{\alpha,1/2}(z). \tag{59}$$

Comparing (59) with (57) Dingle discovered the solutions

$$p(\alpha) \simeq f(h)\, \frac{W_{\alpha,1/2}(8h)}{(\alpha - 1)!}, \qquad g(h)\, \frac{W_{-\alpha,1/2}(-8h)}{(-\alpha - 1)!} \tag{60}$$

where f, g are functions of h which have to be chosen such that

$$p(\alpha) = \sum_i p_i(\alpha) \quad \text{with} \quad p_i \propto h^{-i}.$$

Since for $|z| \to \infty$

$$W_{\alpha,1/2}(z) \sim e^{-(1/2)z} z^\alpha \sum_{s=0}^{\infty} \frac{(s - \alpha)!\, (s - \alpha - 1)!}{(-z)^s\, s!\, (-\alpha)!\, (-\alpha - 1)!}$$

we have

$$p\left(\alpha = j + \frac{1}{2} q\right)$$

$$\propto \frac{(8h)^{-(1/2)q}\, e^{4h}}{\left(j + \frac{1}{2} q\right)!}\, W_{j+(1/2)q,1/2}(8h) \quad \text{for} \quad j < 0$$

and

$$\frac{(-8h)^{(1/2)q}\, e^{-4h}}{\left(-j-\frac{1}{2}q\right)!} W_{-j-(1/2)q,1/2}(-8h) \quad \text{for } j > 0. \tag{61}$$

Picking out the term in h^{-i} and using $(z-1)!\,(-z)! = \pi/\sin \pi z$, we obtain

$$p_i\left(\alpha = j + \frac{1}{2}q\right) \propto \frac{\left(i+\frac{1}{2}q\right)!\left(i+\frac{1}{2}q-1\right)!}{(8h)^i\,(i-j)!\left(j+\frac{1}{2}q\right)!} \quad \text{for} \quad j > 0 \tag{62}$$

and a similar expression for $j < 0$ with $h \to -h$, $q \to -q$, $j \to -j$. Using (54) we now obtain

$$\begin{aligned} P_i(q,j) &\sim 2^{4i} \frac{\left(i+\frac{1}{2}q\right)!\left(i+\frac{1}{2}q-1\right)!}{(i-j)!\left(j+\frac{1}{2}q+1\right)!} \\ &= 2^{4i} \frac{\left(i+\frac{1}{2}q\right)!\left(i+\frac{1}{2}q-1\right)!}{i!} \cdot \frac{i!}{(i-j)!\left(j+\frac{1}{2}q+1\right)!} \\ &\simeq 2^{4i} \frac{\left(i+\frac{1}{2}q\right)!\left(i+\frac{1}{2}q-1\right)!}{i!} \binom{i}{j}. \end{aligned} \tag{63}$$

Hence

$$\begin{aligned} M_{2r} &= 2\sum_{j=1}^{r} j[P_r(q,j)]^2 \\ &\simeq 2^{8r} \left[\frac{\left(r+\frac{1}{2}q\right)!\left(r+\frac{1}{2}q-1\right)!}{r!}\right]^2 \cdot \sum_{j=1}^{r} j \binom{r}{j}^2. \end{aligned} \tag{64}$$

Using

$$\Gamma(2z) = \frac{2^{2z-(1/2)}}{(2\pi)^{1/2}}\,\Gamma(z)\,\Gamma\left(z+\frac{1}{2}\right)$$

and [11]

$$\sum_{j=1}^{r} t\binom{r}{t}^2 \sim \frac{(2r)!}{(r!)^2} \sim 2^{2r} \tag{65}$$

we obtain

$$M_{2r} \sim 2^{8r}(2r+q)!$$

i.e. $\quad M_r \sim 2^{4r}(r+q)!$

and

$$A_r \propto \frac{M_{r+1}}{(2^7)^r} \sim \frac{(r+q)!}{2^{3r}} \tag{66}$$

in agreement with (47).

6. Large-Order Behaviour of Perturbation Expansion for the Anharmonic Oscillator

The case of the periodic potential possesses some unusual simplicity, e.g. the recurrence relation of the coefficients P_i involves only three P_{i-1}'s. In general the calculations are more complicated. As an example we consider [14]

$$\varphi'' + [-L^2 + v(x)]\,\varphi = 0. \tag{67}$$

We expand $v(x)$ around an extremum at x_0:

$$v(x) = v(x_0) + \sum_{i=2}^{\infty} \frac{(x-x_0)^i}{i!}\, v^{(i)}(x_0)$$

and set

$$\left.\begin{aligned} &h = [-2v^{(2)}(x_0)]^{1/2}, \qquad \omega = h(x-x_0) \\ &-L^2 + v(x_0) = \frac{1}{2}\, qh^2 + \Delta, \\ &q \simeq 2n+1, \qquad n = 0, 1, 2, \ldots \end{aligned}\right\} \tag{68}$$

We then have the perturbation expansion [14]

$$\varphi(\omega) = \varphi_q(\omega) + \sum_{i=3}^{\infty} \sum_{j=2i,2i-4,\ldots}^{-2i} P_i(q, q+j)\, \varphi_{q+j}(\omega) \tag{69}$$

where $\varphi_q(\omega) = D_{(1/2)(q-1)}(\omega)$. One can show [12] that the coefficients P_i satisfy the recurrence relation

$$tP_{r-2}(q, q+t) = \sum_{i=3}^{\infty} \sum_{j=2i,2i-4,\ldots}^{-2i} \frac{1}{h^{i-2}}\, P_{r-i}(q, q+j+t) \cdot [q+j+t, q+t]_{i-2} \tag{70}$$

with certain boundary conditions (e.g. $P_{-r} = 0$ for $r > 0$) and

$$\begin{aligned} &[q, q]_1 = \frac{2\Delta}{h} - \frac{v^{(3)}(x_0)}{v^{(2)}(x_0)}\, S_3(q, 0)/3! \\ &i \geqq 3,\ -2i \leqq j \leqq 2i\colon [q, q+j]_{i-2} = -\frac{v^{(i)}(x_0)}{v^{(2)}(x_0)} \cdot \frac{S_i(q, j)}{i!}. \end{aligned} \tag{71}$$

Here the coefficients S_i are given by

$$\omega^i \varphi_q = \sum_{j=2i,2i-4,\ldots}^{-2i} S_i(a, j)\, \varphi_{q+j}. \tag{72}$$

Again we have

$$\frac{2\Delta}{h} = -\sum_{i=2}^{\infty} M^{(i)}$$

where

$$M^{(i)} = \frac{1}{h^i}\, [q, q]_i + \sum_{j=1}^{i-1} \sum_{t \neq 0} tP_{i-j}(q, q+t)\, Q_j(q+t, q) \tag{73}$$

and the coefficients Q_j are defined by an equation similar to (70):

$$tQ_{s-2}(q+t,q) = \sum_{i=0}^{\infty} \sum_{j=2i,2i-4,\ldots}^{-2i} \frac{1}{h^{i-2}} [q+t, q+t+j]_{i-2} \cdot Q_{s-i}(q+t+j, q). \tag{74}$$

The question is therefore: how can we determine P_r (and similarly Q_s)? The main steps are the following. If $v^{(3)}(x_0) \neq 0$ and if $v^{(3)}(x_0)$ constitutes the dominant perturbation we can approximate (70) for large r by

$$thP_{r-2}(q, q+t) \simeq \sum_{j=\pm 2,\pm 6} P_{r-3}(q, q+j+t) \cdot [q+j+t, q=t]_1$$

i.e.

$$-3!\, t \frac{v^{(2)}(x_0)}{v^{(3)}(x_0)} hP_{r-2}(q, q+t) \simeq \sum_{j=\pm 2,\pm 6} P_{r-3}(q, q+j+t)\, S_3(q+j+t, -j). \tag{75}$$

We now define

$$p_{r-i}(q+j+t) \equiv (t+j)^{r-i} \frac{S_i(q+j+t, -j)}{S_i(q+t, j)} \cdot P_{r-i}(q, q+j+t) \tag{76}$$

and consider $|t| \gg |j|$.

Writing

$$p(q+j+t) = \sum_{r+i>0} p_{r+i}(q+j+t) \tag{77}$$

where

$$p_{r+i}(q+j+t) \propto \frac{1}{h^{r+i-2}}$$

the equation for p_{r-i} follows by equating terms of $o(1/h^{r-5})$ on both sides of

$$-3! \frac{v^{(2)}(x_0)}{v^{(3)}(x_0)} hp(q+t) \simeq \sum_{j=\pm 2,\pm 6} S_3(q+t, j)\, p(q+j+t). \tag{78}$$

Comparing (78) with the recurrence relation of parabolic cylinder functions, i.e.

$$\omega^3 \varphi_{q+t}(\omega) = \sum_{j=\pm 2,\pm 6} S_3(q+t, j)\, \varphi_{q+j+t}(\omega)$$

we see that

$$p(q+t) \propto \varphi_{q+t}(\omega)$$

with

$$\omega = -3! \left(\frac{v^{(2)}(x_0)}{v^{(3)}(x_0)} \right)^{1/3} \tag{79}$$

From here on the procedure is similar to that of the previous case. The result is

$$M^{(i)} \sim \left(\frac{3}{2} i \right)! \Big/ h^{i-4}. \tag{80}$$

Thus the terms of the eigenvalue expansion again have the behaviour typical of an asymptotic expansion. For an investigation of the corresponding large-order behaviour for the anharmonic oscillator $ax^2 + bx^4$ we refer to ref. [15].

References

[1] R. B. DINGLE, The role of difference equations, University of St. Andrews preprint, 1964, unpublished.
[2] R. B. DINGLE, Asymptotic expansions: their derivation and interpretation, Academic Press, London 1973.
[3] V. S. POPOV and V. M. WEINBERG, ITEP (Moscow) preprint no. 101 (1982), unpublished.
[4] V. L. ELETSKY, V. S. POPOV and V. M. WEINBERG, Phys. Letters **84 A**, 235 (1981).
[5] H. M. M. MANSOUR and H. J. W. MÜLLER-KIRSTEN, Phys. Rev. **D 24**, 1033 (1981).
[6] R. B. DINGLE, and H. J. W. MÜLLER, J. reine angew. Math. **211**, 11 (1962).
[7] H. J. W. MÜLLER-KIRSTEN, Phys. Rev. **D 25**, 1552 (1982).
[8] H. M. M. MANSOUR and H. J. W. MÜLLER-KIRSTEN, J. Math. Phys. **23**, 1835 (1982).
[9] M. STONE and J. REEVE, Phys. Rev. **D 18**, 4746 (1978).
[10] E. B. BOGOMOLNY and V. A. FATEYEV, Phys. Letters **71 B**, 93 (1977).
[11] R. B. DINGLE and H. J. W. MÜLLER, J. reine angew. Math. **216**, 123 (1964).
[12] L. K. SHARMA and H. J. W. MÜLLER-KIRSTEN, J. Math. Phys. **23**, 367 (1982).
[13] M. ABRAMOWITZ and I. A. STEGUN, Handbook of Mathematical Functions, Dover Publication, New York, 1970, formula 13.4.31.
[14] H. J. W. MÜLLER-KIRSTEN, Phys. Rev. **D 22**, 1962 (1980).
[15] P. ACHUTHAN and H. J. W. MÜLLER-KIRSTEN, University of Kaiserslautern report, 1986, unpublished.

The journal "Fortschritte der Physik/Progress of Physics" is devoted to the theoretical and experimental study of the fundamental constituents of matter and their interactions, so to elementary particle physics, classical and quantum field theory, theory of gravitation, thermodynamics and statistics, nuclear physics, laser physics, and plasma physics.
The articles should have in general review character.

Manuscript should be addressed to the Editor-in-Chief of the journal

Prof. Dr. F. Kaschluhn
Sektion Physik der Humboldt-Universität zu Berlin
Berlin, DDR - 1086, PSF 1297

or to editors or to members of the Advisory Board.

Terms of subscription for the journal "Fortschritte der Physik/Progress of Physics"

Orders can be sent

- *in the GDR:* to the Postzeitungsvertrieb or to the Akademie-Verlag Berlin, Leipziger Str. 3—4, Berlin, DDR - 1086, PF-Nr. 1233;
- *in the other socialist countries:* to a book-shop for foreign language literature or to the competent news-distributing agency;
- *in the FRG and Berlin (West):* to a book-shop or to the wholesale distributing agency Kunst und Wissen, Erich Bieber OHG, Wilhelmstr. 4—6, D - 7000 Stuttgart 1;
- *in the other Western European Countries:* to Kunst und Wissen, Erich Bieber GmbH, General Wille-Str. 4, CH - 8002 Zürich;
- *in other countries:* to the international book- and journal-selling trade, to Buchexport, Volkseigener Außenhandelsbetrieb der DDR, Postfach 160, Leipzig, DDR - 7010 or to the Akademie-Verlag Berlin, Leipziger Str. 3—4, Berlin, DDR - 1086, PF-Nr. 1233.

Zeitschrift „Fortschritte der Physik“

Herausgeber: Prof. Dr. F. Kaschluhn, Prof. A. Lösche, Prof. Dr. R. Rompe; Berlin.
Verlag: Akademie-Verlag Berlin, Leipziger Str. 3—4, Berlin, DDR - 1086;
Fernruf: 2236201 und 2236229; Telex-Nr.: 114420; Bank: Staatsbank der DDR, Berlin, Konto-Nr.: 6836-26-20712.

Chefredakteur: Dr. Lutz Rothkirch.
Anschrift der Redaktion: Sektion Physik der Humboldt-Universität zu Berlin, Invalidenstr. 110, Berlin, DDR - 1040.

Veröffentlicht unter der Lizenznummer 1324 des Presseamtes beim Vorsitzenden des Ministerrates der Deutschen Demokratischen Republik.

Gesamtherstellung: VEB Druckhaus „Maxim Gorki“, Carl-von-Ossietzky-Straße 30/31, Altenburg, DDR - 7400.

Erscheinungsweise: Die Zeitschrift „Fortschritte der Physik“ erscheint monatlich. Die 12 Hefte eines Jahres bilden einen Band. Bezugspreis je Band 300,— DM zuzüglich Versandspesen. Preis je Heft 25,— DM. Der gültige Jahresbezugspreis für die DDR ist der Postzeitungsliste zu entnehmen.

Bestellnummer dieses Heftes: 1027/34/11.

AN (EDV) 48930
01400

Contents of the following issues:

L. JENKOVSZKY: Phenomenology of Elastic Hadron Diffraction

P. BHATTACHARJEE: Do 'Quark-Stars' exist?

O. BERTOLAMI: Brans-Dicke Cosmology with a Scalar Field Dependent Cosmological Term

D. E. B. MANOUKIAN: Charged Particle Emission and Detection Sources and the One-Body and the Many-Body Coulomb Scattering

V. ALDAYA, J. A. DE AZCÁRRAGA: Group Foundations of Quantum and Classical Dynamics: Towards a Globalization and Classification of Some of Their Structures

F. STEINER: Spectrum Sum Rules for the Circular Aharonov-Bohm Quantum Billiard

A. CHAKRABARTI: Classical Solution of Yang-Mills-Fields

K. SVOZIL: Renormalization of the Quantum Theory of the Sölid State

P. D. GALL: Structure of Weak Currents

F. WRESZINSKI: Charges and Symmetries in Quantum Theories without Locality

Progress of Physics is indexed
in Current Contents/Physical, Chemical & Earth Sciences.

FSC
www.fsc.org
MIX
Papier | Fördert gute Waldnutzung
FSC® C083411

Zeitfracht Medien GmbH
Ferdinand-Jühlke-Straße 7
99095 Erfurt, Deutschland
produktsicherheit@kolibri360.de